ADVANCES IN CARBON NANOSTRUCTURES

Edited by **Adrián M.T. Silva**
and **Sónia A.C. Carabineiro**

Advances in Carbon Nanostructures

http://dx.doi.org/10.5772/61730
Edited by Adrián M.T. Silva and Sónia A.C. Carabineiro

Contributors

Mohammad Ali Ali Mohammad, He Tian, Yi Yang, Tian-Ling Ren, Wen-Tian Mi, Irena Kratochvílová, Boris Kharisov, Oxana V. Kharissova, Beatriz Ortega García, Ubaldo Ortiz Mendez, Saiful Izwan Abd Razak, Nadirul Hasraf Mat Nayan, Nor Syuhada Azmi, Abdul Halim Mohd Yusof, Izzati Fatimah Wahab, Farah Nuuruljannah Dahli, Noor Fadzliana Ahmad Sharif, Youngduck Kim, Myung-Ho Bae, Quan Xu, Wenwen Zhang, Liu Shuan, Hammad Younes, Vesselin Shanov, Seyram Gbordzoe, Rachit Malik, Noe Alvarez, Robert Wolf, Davide Micheli, Roberto Pastore, Antonio Vricella, Mario Marchetti

CBS, Edition **2017**

Published by InTech

Janeza Trdine 9, 51000 Rijeka, Croatia

Publishing Process Manager Romina Rovan

Technical Editor SPi Global

Cover Designer InTech Design Team

Additional hard copies can be obtained from orders@intechopen.com

Advances in Carbon Nanostructures, Edited by Adrián M.T. Silva and Sónia A.C. Carabineiro
p. cm.
Print ISBN 978-953-51-2642-3
Online ISBN 978-953-51-2643-0

INTECH

open science | open minds

Contents

Preface

Carbon materials are outstanding multipurpose resources that can show several forms. Carbon atoms have the ability to bond in different manners that can assume distinct dimensional arrangements from which diverse nanostructures are obtained. Graphene and its derivatives, carbon nanotubes, and diamonds are among the most recently investigated carbon materials. These carbon materials are very versatile and can be easily tuned to obtain samples with different properties that can fit several various applications. This book aims to cover the topic of "Advances in Carbon Nanostructures" and includes 12 book chapters. The first nine chapters consist of significant reviews dealing with diverse carbon-based configurations and applications, while the last three chapters show some case studies in the field. These chapters are organized in three sections: (i) *Graphene and derivatives,* (ii) *Carbon nanotubes and diamonds,* and (iii) *Nanofluids and case studies on carbon nanostructures.*

The first section deals with *Graphene and derivatives.* **Chapter 1** describes the preparation, properties, and applications of magnetic graphene-based nanocomposites, including materials based on elemental metals (such as Fe, Co, and Ni), magnetic nanoclusters, various morphological forms of iron oxides, ferrites, 3D graphene aerogels on Fe_3O_4 nanoclusters, single-molecule magnets, and organometallic and polycomponent composites, among others. **Chapter 2** includes the latest advances on graphene-based membranes for gas separation and water purification, aiming to understand details involved in the molecular transport through these membranes and discuss the recent challenges, as well as the expected developments in this topic. **Chapter 3** focuses on novel devices covering electric, acoustic, photonic, magnetic, and mechanical domains that can be obtained with graphene and introduces the laser-scribing technology, which enables wafer scale production of graphene devices. **Chapter 4** deals with light emission from graphene, a subject not yet fully explored, but with the potential to open the door toward obtaining atomically thin, flexible, and transparent light sources and graphene-based on-chip interconnects. **Chapter 5**, the last of the first section, discusses the work carried out on electrospinning of graphene oxide-based nanofiber materials, focusing on the combination of graphene oxide with polymeric materials, as well as their properties and applications.

The second section gathers the chapters dealing with *Carbon nanotubes and diamonds.* **Chapter 6** reports the design and fabrication of low-voltage carbon nanotube heaters and illustrates their potential applications by the manufacturing and testing of heatable gloves and via deicing experiments. **Chapter 7**summarizes recent accomplishments in the reinforcement of cement using carbon nanotubes and fibers, with emphasis on several effects upon dispersion of carbon allotropes in concrete-water media along with their possible health impact. **Chapter 8** reviews the technological achievements in the synthesis of diamond thin films of the last decade, which led to the use of the remarkable diamond properties and to the development of a wide range of applications in various fields of engineering.

Finally, the third section covers the *nanofluids and case studies on carbon nanostructures.* **Chapter 9** reviews recent studies dealing with the use of both graphene and carbon nanotubes to improve the thermal conductivity of nanofluids (i.e., liquid suspensions of nanoparticles, which have higher thermal conductivity than the base fluids), highlighting the effect of using magnetic fields to enhance their thermal conductivity. The last three chapters focus on specific case studies. In particular, **Chapter 10** deals with the fabrication and characterization of scale sensors using casting and inkjet printing of carbon nanotubes. **Chapter 11** reports the protective mechanism, wear resistance, and antifriction properties of graphene-based epoxy coatings used in marine environment. **Chapter 12** refers to fully configurable electromagnetic wave absorbers employing carbon nanostructures, a new method for the full design and configuration of layered carbon-based nanocomposite materials able to perform the quasi-perfect imitation of any type of electromagnetic reflection coefficient target profile.

In summary, these book chapters present several fundamental concepts, recent advances, and future challenges in the synthesis and technological application of three target groups of nanostructured carbon materials. For this reason, this book is expected to stimulate the worldwide community working in this field, and the editors trust that readers will find this book interesting and helpful. The editors are extremely grateful to all authors, affiliated in four different continents, for their willingness to contribute to this book as well as for meeting the deadlines on time. An effort has been done to avoid overlapping of content between different chapters, but some differences in writing and style result from the large information of the many different contributions received. The editors also thank their main funding agencies (FCT—Fundação para a Ciência e Tecnologia through IF/01501/2013 and IF/01381/2013/CP1160/CT0007 with financing from the European Social Fund and the Human Potential Operational Programme and project POCI-01-0145-FEDER-006984 through FEDER, COMPETE2020, and POCI) as well as the exceptional assistance of the InTech staff members, especially Ms. Romina Rovan, Ms. Ana Simčić, and Ms. Petra Nenadic, for their patience and unflagging support in all stages of the editorial process.

Adrián M.T. Silva and Sónia A.C. Carabineiro
Principal Associate Researchers
Laboratory of Separation and Reaction Engineering—Laboratory of Catalysis and Materials (LSRE-LCM)
Faculdade de Engenharia, Universidade do Porto
Porto, Portugal

Section 1

Graphene and Derivatives

Chapter 1

Magnetic-Graphene-Based Nanocomposites and Respective Applications

Oxana Vasilievna Kharissova, Beatriz Ortega García,
Boris Ildusovich Kharisov and Ubaldo Ortiz Méndez

Additional information is available at the end of the chapter

http://dx.doi.org/10.5772/64319

Abstract

Preparation, properties, and applications of magnetic-graphene-based nanocomposites are reviewed. Graphene magnetic nanocomposites include those on the basis of elemental metals (Fe, Co, Ni), magnetic nanoclusters, various morphological forms of iron oxides (Fe_2O_3, Fe_3O_4), ferrites MFe_2O_4, 3D graphene aerogels@hierarchical Fe_3O_4 nanoclusters, single-molecule magnets like $TbPc_2$ (Pc: phthalocyanine), other organometallic-containing composites (benzene-metal-graphene), as well as polycomponent nanocomposites such as Ag/Fe_3O_4/G (G: graphene), Fe_3O_4/CdS/G, or FePc/Fe_3O_4/GO (GO: graphene oxide), among others. Their available synthesis methods consist commonly of hydrothermal and solvothermal techniques, sol-gel autocombustion, sonoelectrochemical polymerization, thermal expansion and thermal reduction, microwave-assisted technique, and covalent bonding chemical methods. Their current and potential applications are distinct devices, in particular for colorimetric detection of glucose, construction materials, analytical, sensor and biosensor applications, environmental remediation, compounds with antibacterial properties, catalysis and photocatalysis, biological imaging, oil absorption, etc.

Keywords: environmental applications, ferrites, grapheme, hybrids, iron oxides, lithium batteries, magnetic composites, metals

1. Introduction

Graphene, which consists in a single layer of carbon atoms, has been intensively studied in the last decade due to its extraordinary structural and electronic properties, leading to numerous current and potential applications in electronic devices. Growing, doping, or adsorption of metal nanoparticles and other inorganic/organic species on its surface can change, improve, and manipulate graphene magnetic and electronic properties. If these particles possess magnetic properties, the whole nanocomposite can have them too, opening new opportunities and applications for the formed hybrids.

Currently, the information on magnetic graphene and graphene oxide nanocomposites has not yet been extensively reviewed despite a number of original reports. At the same time, their importance is undoubtful due to numerous current and potential applications of graphene itself, and magnetic nanoparticles in a variety of fields of chemistry, physics, medicine and biology, in the frontiers with nanotechnology. This topic is rarely presented as short sections in some reviews [1–4], dedicated to graphene, in contrast to carbon nanotubes, where a concise review was recently published [5]. So, the main purpose of this chapter is to highlight magnetic graphene composites, and review their main synthesis methods, properties, and current and potential applications.

2. Metal-graphene composites

A certain number of magnetic metals are present in the Periodic Table. Graphene composites have been reported only with some typical metals with paramagnetic properties, such as iron, cobalt, or nickel. Among theoretical studies of metal interactions with graphene surface, quantum Monte Carlo methods, studying the consequences of placing a magnetic adatom adjacent to a vacancy in a graphene sheet [6, 7], and description of electronic and magnetic properties of the graphene-ferromagnet interfaces [8] are particularly relevant. In addition, the plane-wave density functional theory (DFT) was used to study the properties of nanostructures of the type C_{60}-M-G (M: Ti, Cr, Mn, Fe, or Ni; G: graphene) [9]. In this work, in particular, it was found that the high-spin C_{60}-Cr-G nanostructure is more stable than its low spin analog. Structures containing distinct metals can be different. For instance, on the graphene surface, C_{60}-Ti stands symmetrically upright, while for other metals, C_{60}-M fragments are bent and nonequal in each composite (**Figure 1**). Authors suggested, based on calculations, that the C_{60}-M fragment is flexible, respecting the bending motion. Importantly, the attachment of C_{60} to an M-G surface is more energetically favored than the decoration of graphene with C_{60}-metal complexes. DFT studies were also carried out for Co-G composites (cobalt-vacancy defect [10] or single Co atom incorporated with divacancy in armchair graphene nanoribbon [11]). In this case, metal dopation was found to introduce magnetic properties into the formed composite.

Several reports were dedicated to *cobalt-graphene* nanocomposites. Studying single Co atoms on graphene on Pt(111) and their magnetic properties by scanning tunneling microscopy spin-excitation spectroscopy, upon hydrogen adsorption, three differently hydrogenated species were identified (**Figure 2**) [12], whose magnetic properties are very different from those of cobalt element. It was suggested that, due to the interaction of Co-G orbitals, large magnetic anisotropy stems from strong ligand field effects that take place. A series of reports studied the *intercalation* of cobalt between graphene and Ir(111) via distinct techniques. Thus, cobalt layers intercalated in between graphene and Ir(111) were characterized by a combination of

X-ray magnetic circular dichroism and photoemission [13]. In particular, an induced magnetic moment in the graphene layer, antiparallelly oriented to that of cobalt, was established. Magnetic ordering was found to exist beyond monolayer intercalation. In a related report [14], cobalt-intercalated graphene on Ir(111) was studied by DFT calculations and spin-polarized scanning tunneling microscopy. It was revealed that the strong bonding between cobalt and graphene leads to a high corrugation within the Moiré pattern, which arises due to the lattice mismatch between the graphene and the Co on Ir(111).

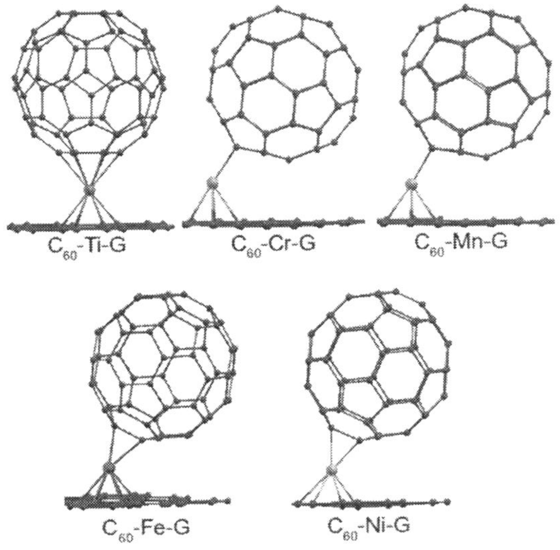

Figure 1. C_{60}-M-G nanostructures. Reproduced with permission form the *American Chemical Society*.

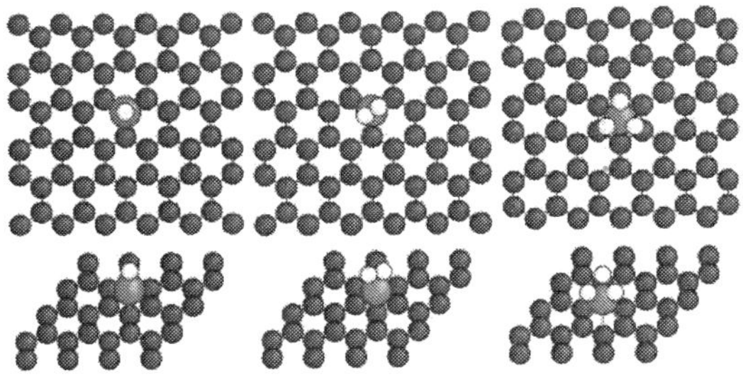

Figure 2. Top and side view of atomic structures of CoH_n (n = 1, 2, and 3) complexes. Reproduced with permission from the *APS Physics*.

In addition, nanometer-sizes films of cobalt were intercalated at the interface "graphene/ Ir(111)" and were characterized by Auger electron spectroscopy and spin-polarized low-energy electron microscopy [15]. In resulting composites, graphene top layer was found to promote perpendicular magnetic anisotropy in the Co film. It was also revealed that the magnetic anisotropy energy is significantly larger for the graphene/Co interface than for the free metallic surface. Regarding metal intercalation, contrarily to cobalt, Fe-based intercalation of GO led to the formation of reduced GO/Fe_3C magnetic hybrids [16]. Iron carbide particles were encapsulated in a graphite cage, protecting them from agglomeration.

Among other available theoretical studies on Co/G nanocomposites, the single Co layer added on graphene showed ferromagnetic ordering, with perpendicular alignment to the graphene sheet [17], according to the data obtained by relativistic density functional theory, at the level of generalized gradient approximation. These properties are promising for magnetism in 2D systems despite experimental difficulties in obtaining regular structures of single layer of this metal on graphene. In addition, the oxidation state of cobalt, together with possible oxidation of metal in real conditions, is also important. Thus, XPS spectral data for graphene/Co composites, obtained from $CoCl_2$. $6H_2O$ diluted in ethyl alcohol and highly split graphite, showed two sets of $2p_{3/2,1/2}$ lines belonging to partly oxidized and metallic Co atoms [18]. The formation of this protective oxide layer prevents metallic cobalt from deep oxidation.

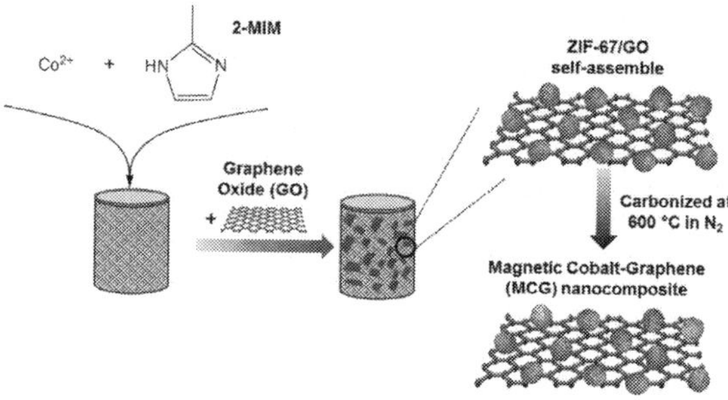

Figure 3. Scheme of synthesis of magnetic cobalt–graphene (MCG) nanocomposite. Reproduced with permission from the *Royal Society of Chemistry*.

The synthesis methods for cobalt-graphene nanocomposites frequently include high-temperature decomposition of precursors. Thus, a magnetic cobalt-G nanocomposite (**Figure 3**) was prepared by carbonizing a self-assembly of a cobalt-based metal organic framework, ZIF-67, and GO [19]. This composite based on cobalt and reduced graphene oxide (RGO) was used as a catalyst for the activation of peroxymonosulfate (PMS) in the process of decolorizing Acid Yellow dye in water. Resulting regeneration efficiency remained at 97.6% over 50 cycles, showing its effective and stable catalytic activity. The proposed mechanism of MCG activating PMS to generate sulfate radicals is shown in **Figure 4**. Another example is a combination of autocombustion and sol-gel methods, which led to Co-G nanocomposite, prepared from GO, $Co(NO_3)_2$, and citric acid as precursors [20]. A sol-gel was first prepared and then underwent autocombustion in Ar atmosphere, at 300°C, due to the action of the produced reducing agents H_2 and CH_4. In the formed nanocomposite, Co nanoparticles (with a diameter of about 10 nm) are homogeneously distributed on graphene surface. The same technique can be applied for loading Ni, Cu, Ag, and Bi on graphene surface. Pyrolisis was also used, for instance, in the case of cobalt phthalocyanine (CoPc), for the purpose of fabricating organic metal/graphene composites [21]. CoPc is capable of dispersing cobalt and its oxide onto graphene sheets due to π-interactions between them in the conditions of pyrolysis or oxidation. The process of the fabrication of Co/GC and Co_3O_4/GC is shown in **Figure 5**. The Co_3O_4/graphene nanocomposites showed remarkable lithium storage performance, including good rate capability, good cycle performance, and highly reversible capacity. It was suggested that CuO, Fe_2O_3 and other metal or metal-oxide-based graphene composites can also be prepared in this way.

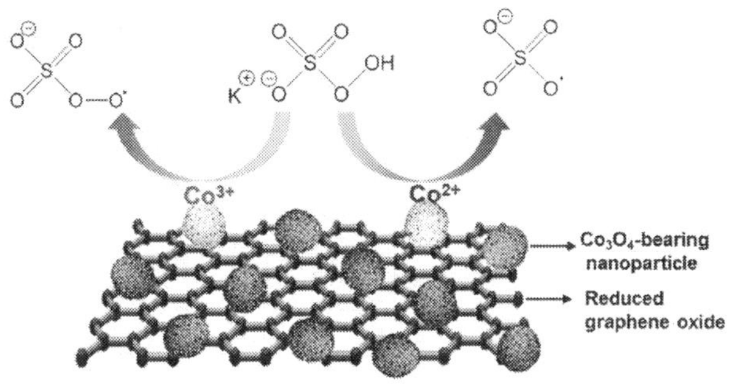

Figure 4. Proposed mechanism for PMS activation generating sulfate radicals. Reproduced with permission from the *Royal Society of Chemistry*.

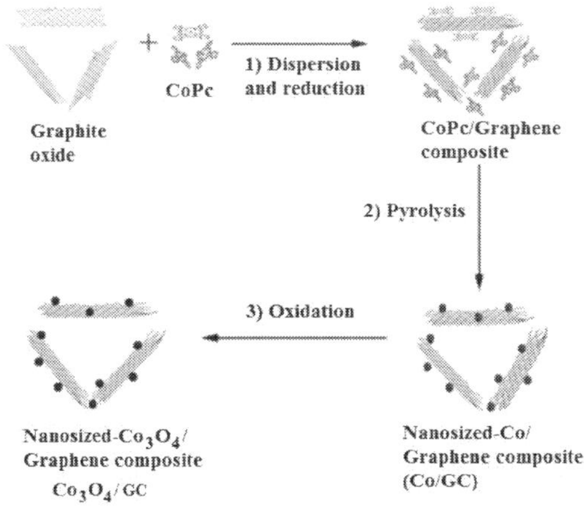

Figure 5. Fabrication of Co/GC and Co_3O_4/GC: (1) dispersion of graphite oxide and CoPc in water by ultrasonication and subsequently chemical reduction of graphite oxide, (2) pyrolysis at 800°C, and (3) oxidation at 400°C. Reproduced with permission from the *John Wiley & Sons*.

The decolorization of organic dyes, such as the Acid Yellow as mentioned above, is one of the classical applications of graphene composites of cobalt and other magnetic metals. Cobalt nanoparticles (about 30 nm size), anchored on graphene sheets, were tested for heterogeneous oxidation of a dyeing pollutant, Orange II, with peroxymonosulfate (PMS) in aqueous solutions [22]. In comparison to pure cobalt, the incorporation of Co nanoparticles into graphene sheets resulted in much higher catalytic activity for Orange II degradation. The proposed reaction mechanism is described by reactions (1)–(7). This work shows the importance of using other metal nanoparticles rather than iron for the decontamination of organic pollutants.

$Co^0 \rightarrow Co^{2+} + 2e^-$ *(1)*

$\equiv Co^0 + O_2 + 2H_2O \rightarrow \equiv Co^{2+} + 4OH^-$ *(2)*

$\equiv Co^0 + 2HSO_5^- \rightarrow \equiv Co^{2+} + 2SO_4^{-\cdot} + 2OH^-$ *(3)*

$\equiv Co^{2+} + HSO_5^- \rightarrow \equiv Co^{3+} + SO_4^{-\cdot} + OH^-$ *(4)*

$\equiv Co^{3+} + HSO_5^- \rightarrow \equiv Co^{2+} + SO_5^{-\cdot} + H^+$ *(5)*

$\equiv Co^0 + 2 \equiv Co^{3+} \rightarrow 3 \equiv Co^{2+}$ *(6)*

$SO_4^{-\cdot}$ + Orange II → many steps → $CO_2 + H_2O$ *(7)*

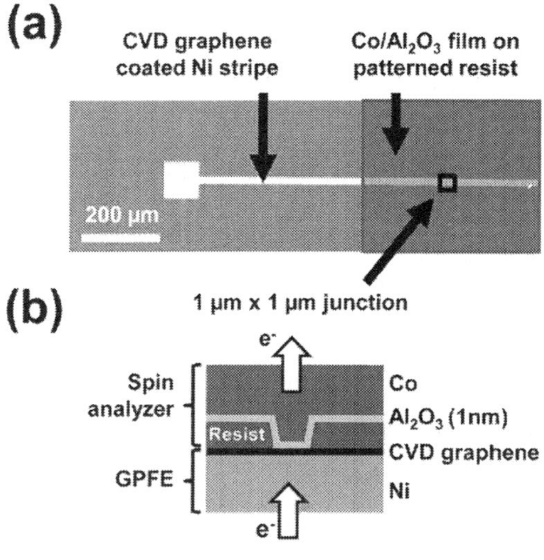

Figure 6. (a) Optical image of the device after graphene growth: an Ni stripe is coated with CVD graphene and an Al_2O_3/Co electrode is then deposited on lithographed squares in UV resist. (b) Cross-sectional scheme of the junction. Reproduced with permission from the *American Chemical Society*.

In the case of *nickel*, its graphene or GO composites are not so widespread compared to cobalt or iron. Nickel itself is used as a support for graphene fabrication, for instance, by the CVD technique of CH_4, leading to low-defect synthesis of bilayer graphene on evaporated polycrystalline nickel films [23]. The CVD method also allows to obtain nickel coated with a few layers of graphene, resulting in a stable surface upon air exposure [24]. This way, graphene-passivated (against oxidation) ferromagnetic electrodes can be fabricated, which are suitable for spin devices (namely spin-polarized oxidation-resistant electrodes as shown in **Figure 6**).

In addition to CVD, graphene or GO hybrids with nickel have also been obtained by other methods, for example, grown on a graphene Moire on Rh(111) at 150 K [25]. These processes of graphene decoration with nickel nanoparticles can include simultaneous reduction of GO in the process of synthesizing GO/Ni composites [26], in particular simultaneous reduction of GO and nickel(II) ions by the one-step far-infrared-assisted method [27]. Also, nickel nanoparticles were loaded by a graphene nanosheet by easily scalable and reproducible direct electrochemical deposition [28]. The graphene sheet decorated with magnetic nickel nanoparticles yielded a composite with soft magnetic and conductive properties, which efficiently promoted microwave absorbability (and was better than graphene alone).

Iron/graphene (or iron/GO) hybrids are very well studied and, together with other iron-containing nanostructures below (iron oxides and their combinations with iron, as $Fe@Fe_2O_3$), are undoubtedly the center of magnetic metal/graphene field. The hybrids can be obtained by several methods, both classic chemical and "greener" routes. Thus, *focused solar radiation*, as a green chemical route, was applied for the fabrication of 3D metal/metal oxide nanoparticles dispersed on 2D ultrathin graphene by a simultaneous reduction and exfoliation process [29]. This procedure allowed the insertion of nanoparticles between ultrathin graphene layers acting as "spacers" between them. These metal/metal oxide nanoparticle dispersed graphene composites might have potential applications in environmental fields, conversion devices, energy storage sensing, and heterogeneous catalysis.

Microwave irradiation was used for the synthesis (**Figure 7**) of combined hierarchical 3D graphene/carbon nanotube/iron nanostructures (G-CNT-Fe) [30]. The formed composite consists of vertically aligned CNTs, which are grown in graphene sheets along with shorter branches of CNTs stemming out from both the vertically aligned CNTs and the graphene sheets. Zero-dimensional (0D) functional iron oxide nanoparticles decorated within this 3D hierarchical nanostructure (both on 1D nanotubes and 2D graphene sheets) provide outstanding lithium storage characteristics. Iron moieties were found to be present in mixed valence states of FeOOH and Fe_2O_3. In addition, iron nanoparticles can be first deposited onto a graphene/Cu substrate by *vacuum deposition* and then the hydrogenation was carried out at 1 atm of gaseous H_2 and under liquid N_2 [31]. Before the experiment, the stabilization of hydrogenated Fe nanoparticles on this support was predicted by DFT calculations, meanwhile the existence of Fe hydride is considered as nonreachable. Hydrogen was found to be released from hydrogenated Fe nanoparticles. In addition, a *thermal reduction* method was used for the synthesis of $Fe/Fe_3O_4/G$ nanocomposite electrode material, using iron oxalate and exfoliated graphene oxide as precursors [32]. Its application, along with nickel, for rechargeable Ni/Fe alkaline batteries (discharge and charge capacities of 280 mAh/g) in hybrid electric vehicles can be applied in a large-scale production.

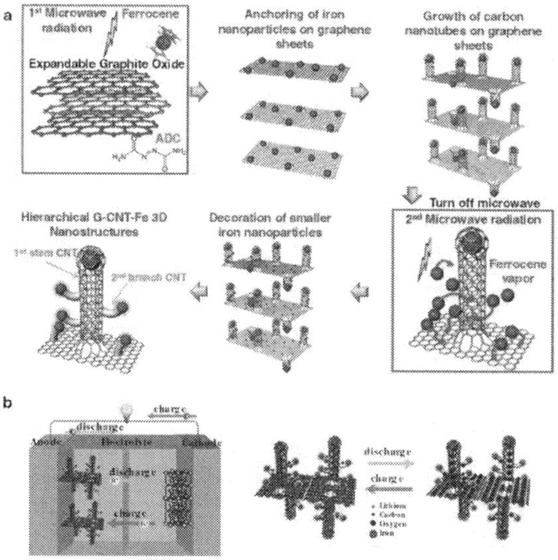

Figure 7. Two-step synthesis of hierarchical G-CNT-Fe 3D nanostructure and its application to anode material in lithium-ion batteries. (a) Step 1: microwave reaction for carbon nanotube growth. Step 2: iron nanoparticle decoration and fast explosive growth of smaller carbon nanotubes on original carbon nanotubes and graphene. (b) Schematic diagram for charge and discharge in G-CNT-Fe 3D anode material. Reproduced with permission from the *American Chemical Society*.

Sodium borohydride has been frequently used as classic reductive agent for the reduction of iron ions and/or GO. Thus, nano zero-valent iron (nZVI) -decorated graphene sheets were prepared via sodium borohydride reduction of GO and applied for Cr(VI) removal [33]. In a related report [34], nonhazardous superparamagnetic nanocomposites, consisting of iron nanoparticles (5 nm) and graphene, were synthesized from GO through intermediate formation of Fe^{3+}/GO complexes and their further reduction with $NaBH_4$. Methyl blue solution, a dye often present in wastewater of dyeing industry, can easily be decolorized using this nanocomposite.

Iron/G (or GO), as well as iron oxide/G (or GO) hybrids, can also have applications in the catalysis area. A novel type of oxygen reduction reaction (ORR) electrocatalyst on the basis of few-walled (2–3 walls) CNT-G complexes was reported [35]. Abundant defects on the outer walls of the CNTs can be produced via partial unzipping of the outer CNTs walls and the formation of large quantities of graphene sheets, connected with the intact inner walls of the CNTs. These graphene sheets make easier the formation of catalytic sites for ORR on annealing in NH_3. Nitrogen doping and Fe impurities are responsible for higher ORR activity in the CNT-G complexes. It was established that Fe atoms are often close to N atoms and are frequently situated along edges of defective graphene sheets. Elimination of Fe by purification leads to a considerable decrease in ORR activity. Indeed, both nitrogen and iron atomic species are important to the high ORR electrocatalytic activity. Another example is a cost-effective synthesis of nitrogen-doped graphene (NG; **Figure 8**) carried out by using cyanamide as a nitrogen source and graphene oxide as a precursor; iron nanoparticles were incorporated into NG using $FeCl_3$ as precursor [36]. This composite was used as a model for the elucidation of the influence of nonnoble metals on the electrocatalytic performance. The NG supported with iron nanoparticles (5 wt%) showed high current density (8.20 mA cm^2) in an alkaline solution and an excellent methanol crossover effect, high stability in distinct media, high surface area, among other advantages, in comparison with Pt and NG-based catalysts, thus allowing platinum replacement.

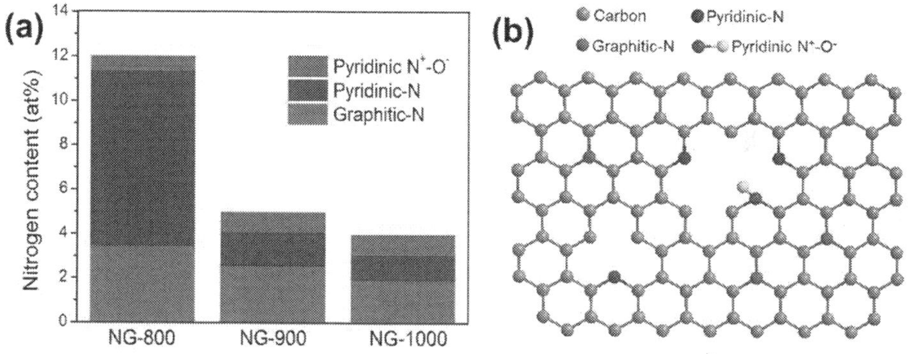

Figure 8. (a) The content of three types of nitrogen in NG. (b) Schematic representation of NG. Reproduced with permission from the *American Chemical Society*.

Environmental problems of contamination with heavy metals and other pollutants (As, Cr) can be partially solved using iron/graphene composites, in particular those containing iron in both metallic and oxidized form. A one-pot thermal decomposition method was used for the preparation of graphene nanoplatelet composites decorated with core-shell Fe-Fe_2O_3 nanoparticles [37]. These nanocomposites could be separated from the liquid-phase mixture with the aid of a permanent magnet. Efficient and effective adsorption of arsenic(III) from the polluted water was observed for this material (nearly complete As(III) removal within 1 ppb) and attributed to the increased adsorption sites existing in the presence of magnetic nanoparticles. Magnetic graphene nanocomposites (MGNCs), consisting of a core@double-shell structure of the nanoparticles with crystalline Fe as the core, iron oxide as the inner shell, and amorphous Si-S-O compound as the outer shell, were prepared by a thermodecomposition process (**Figure 9**) [38]. These composites were highly stable even in 1 M HCl aqueous acid and showed a fast and highly efficient removal of Cr(VI) from wastewater after 5 min (**Figure 10**), in contrast to several other materials (like carbon or waste biomass), the use of which require several hours or days and are not able to achieve 100% removal of Cr(VI).

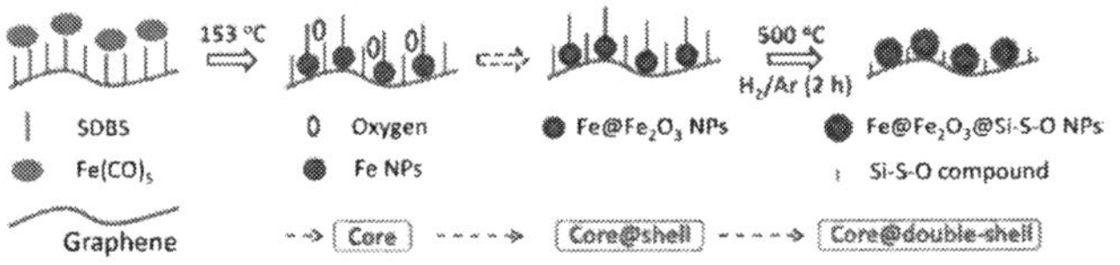

Figure 9. Schematic illustration of the formation of the MGNCs. Reproduced with permission from the *American Chemical Society*.

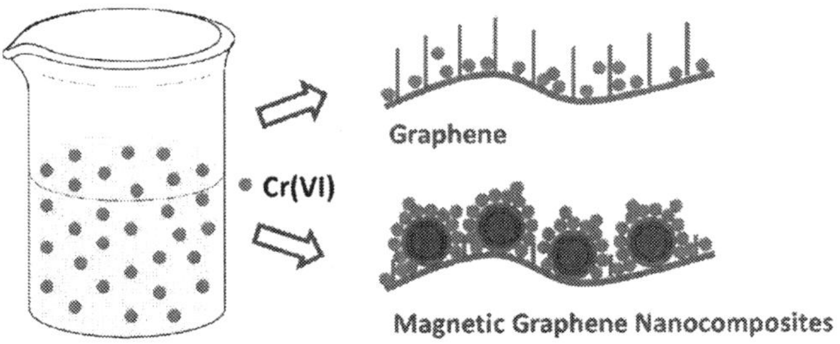

Figure 10. Schematic adsorption mechanisms on graphene and MGNCs. Reproduced with permission from the *American Chemical Society*.

Another application is related to polymers: graphene nanosheets, consisting of iron core and iron oxide shell nanoparticles, G/Fe@Fe_2O_3, were used as nanofillers for the fabrication of magnetic epoxy resin polymer nanocomposites [39]. The G/Fe@Fe_2O_3 was found to favor char formation from the epoxy resin; in addition, the tensile strength of polymer nanocomposite with 1.0 wt% Gr/Fe@Fe_2O_3 was found to be 58% higher than that of the pure epoxy and it was attributed to the high stiffness of graphene. It was suggested that the porous char layer with Gr/Fe@Fe_2O_3 may indicate the existence of Fe@Fe_2O_3, which helps the formation of gas, during the decomposition process of epoxy resin.

3. Fe_2O_3-graphene hybrids

Iron can be present in graphene composites in the elemental form (as shown above) or as Fe core/oxide shell nanoparticles, iron oxides, among others. The ratio of iron nanoforms in *different oxidation states* depends, in particular, on O-containing groups present in the graphene surface, use of reductants, and other conditions. Magnetic iron-containing nanoparticles were loaded on the GO sheets due to the abundant oxygen-containing functionalities present in these carbon materials (hydroxyl, epoxy, and carboxyl functional groups), and their growth mechanism was studied [40]. Most of these functional groups were eliminated and the magnetic nanoparticles were partially converted to iron during thermal treatment under reducing conditions. Metal nanoparticles changed the GO lattice structure and intrinsic functionalities; this effect depended on the amount of iron precursor.

The *effect of pH* is also very important and it was studied for several noncovalent magnetic GO-based materials, prepared using Fe_2O_3 microparticles, nanoparticles, and magnetic surfactants [41]. pH adjustment was used to effectively charge repulsion or attraction between Fe_2O_3 particles and the GO sheets (**Figure 11**). Each material caused coflocculation of GO at acidic pH, leading to materials that could be captured using an external magnetic field. The adsorption of GO at low pH was explained by attractive electrical double-layer forces between the GO and Fe_2O_3 or surfactants. On the contrary, at higher values of pH, the dispersions are stable due to alike-charge repulsions. An intriguing effect was found with Fe_2O_3 nanoparticles: low concentrations resulted in the flocculation of GO and higher concentrations caused restabilization, which are explained by an effective overcharging of the GO surfaces. These systems were found to remove a model nanomaterial, gold nanoparticles, from water.

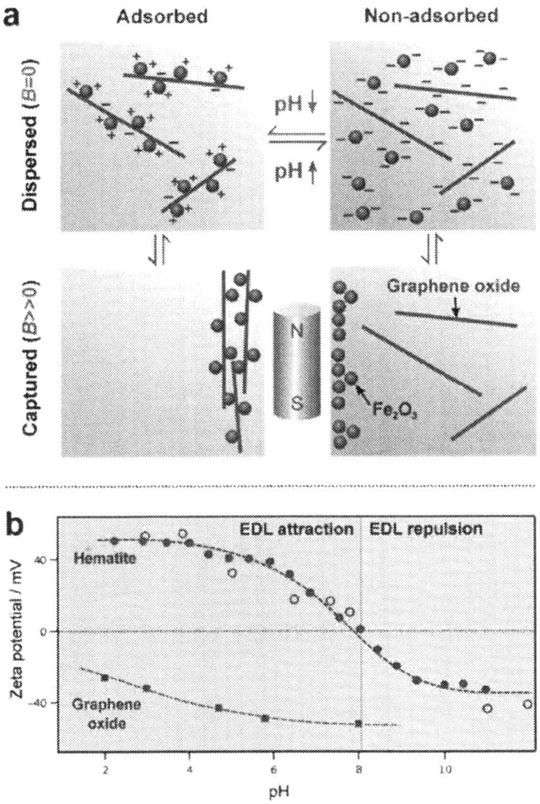

Figure 11. (a) Conceptual scheme of the experiment: pH adjustment is used to affect charge attraction or repulsion between the GO sheets and Fe_2O_3 particles. A magnetic field can be used to separate the Fe_2O_3 from solution or dispersion. (b) Zeta potentials of graphene oxide and Fe_2O_3, demonstrating the pH ranges at which electrical double-layer attraction or repulsion would be expected. Reproduced with permission from the *American Chemical Society*.

The main synthesis methods for iron(III) oxide/graphene nanohybrids are, in general, similar to those used for metals described above or Fe_3O_4 that are explained below. Thus, iron oxide nanoparticles encased by permeable carbon layers of few-layer graphene were synthesized by high-pressure pyrolysis of ferrocene with pristine graphene [42]. The ferrocene precursor provides both carbon and iron, leading to the carbon-coated iron oxide, while the graphene works as a high-surface-area anchor, to obtain small iron oxide nanoparticles. This material was used to improve the electrochemical performance of iron-oxide-based electrodes on Li-ion batteries. Similarly, an iron-oleate precursor (**Figure 12**) was used for the preparation of an iron-oxide/graphene nanocomposite via a solventless thermal decomposition method [43]. Highly monodisperse γ-Fe_2O_3 nanoparticles were found to be in close contact with graphene. This nanomaterial can serve as a potentially valuable candidate anode material for high-rate Li-ion batteries.

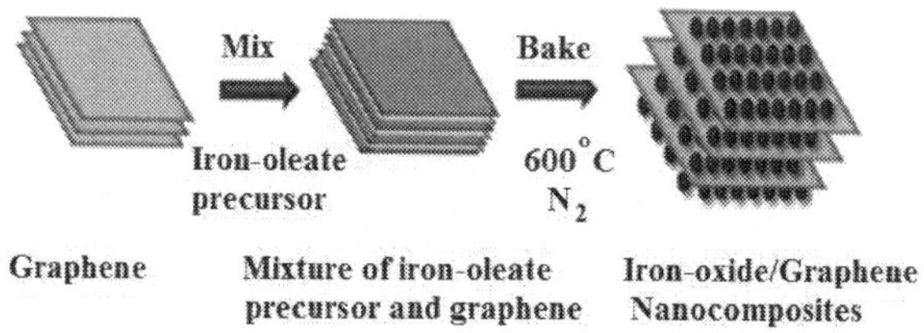

Figure 12. Schematic representation of the direct preparation of iron-oxide/graphene nanocomposites by the solventless thermal decomposition method. Reproduced with permission from the *Royal Society of Chemistry*.

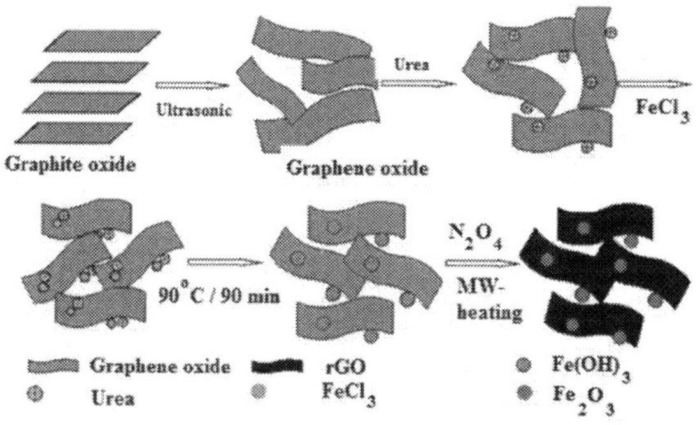

Figure 13. Scheme of the formation of the rGO/Fe_2O_3 composite. Reproduced with permission from the *American Chemical Society*.

In addition to high-temperature destructive methods, sol-gel and coprecipitation techniques are also common. Thus, reduced graphene oxide (rGO) tethered with maghemite (γ-Fe_2O_3) was prepared by a sol-gel process without a reducing agent in which sodium dodecylbenzenesulfonate (NaDDBS) was added into the suspension for the prevention of undesirable formation of an iron oxide 3D network [44]. These composites were applied as anodes for half lithium-ion cells, exhibiting improved cycle life, reversible capacity, and good rate capability. Two-step synthesis (**Figure 13**), consisting of homogeneous precipitation and subsequent microwave-assisted reduction of the GO with hydrazine, led to reduced graphene oxide (rGO) platelets decorated with Fe_2O_3 nanoparticles, uniformly distributed on the surface of platelets [45]. The total specific capacity of rGO/Fe_2O_3 was determined to be higher than the sum of pure rGO and nanoparticular Fe_2O_3. In addition, polypyrrole (PPy) was reinforced with rGO and Fe_2O_3 to reach electrochemical stability and enhancement [46]. This ternary nanocomposite film was fabricated using a one-pot chronoamperometry approach.

In addition to their use in batteries, as referred above, a few other uses are known for Fe_2O_3/G hybrids. Thus, functional nanocomposite-based selective separation of microcystin-LR (toxin belonging to the family of microcystins produced by cyanobacteria and known to be the most toxic of this group) from contaminated water was achieved (**Figure 14**), applying cyclodextrin-functionalized magnetic composites consisting in porous silica and colloidal graphene [47]. In this material, the magnetic component offers easier separation of microcystin-LR from water and the cyclodextrin constituent offers host-guest interaction with microcystin-LR. Among all experimented cyclodextrins, γ-cyclodextrin was found to offer the best performance. The studied functional nanomaterials can be used for the development of advanced water purification systems. Catalytic applications are also reported, for example, for nanosized Fe_2O_3 composites on carbon matrix, modified with nitrogen-doped graphene (**Figure 15**), that are excellent catalysts for the chemoselective hydrogenation of nitroarenes to anilines with good yields (reaction (8)) [48].

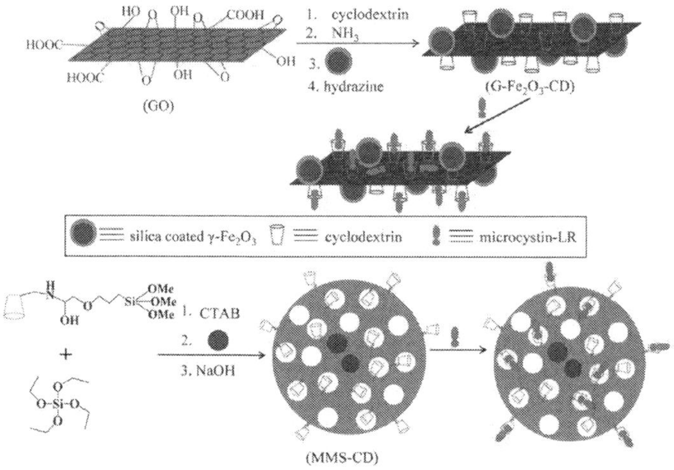

Figure 14. Synthesis strategies for cyclodextrin functionalized magnetic graphene composite (G-Fe_2O_3-CD) and cyclodextrin functionalized magnetic mesoporous silica (MMS-CD). Reproduced with permission from the *American Chemical Society*.

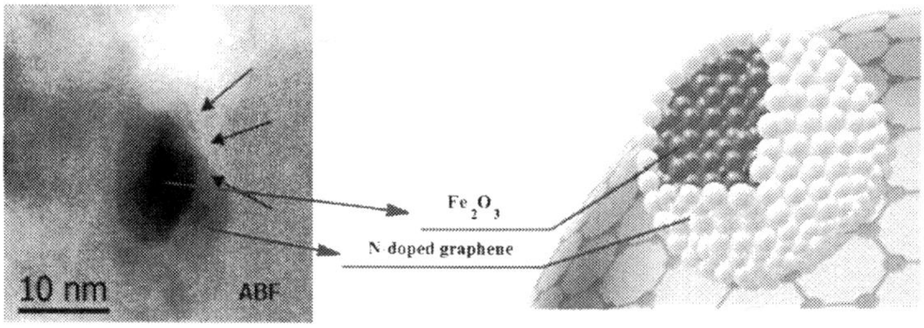

Figure 15. Nanoscaled Fe_2O_3 particles surrounded by N-doped graphene layers. Reproduced with permission from the *American Chemical Society*.

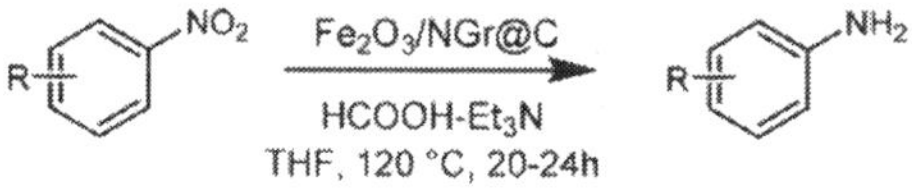

(8)

4. Fe_3O_4-graphene composites

4.1. Fe_3O_4/G aerogels

Among a large variety of mixed-valent-iron-oxide/graphene (or GO) composites, considerable attention is paid to *aerogels* in contrast to the above-described magnetic graphene hybrids. Graphene and other aerogels and routes for their synthesis are well known [49]. The introduction of a magnetic component could lead to a larger variety of unusual properties and potential applications, namely, the easy removal of pollutants, such as crude oil. These nanocomposites can have a simple composition (i.e., Fe_3O_4/G) or contain an additional component like a polymer. Thus, 3D N-doped graphene aerogel (N-GA)-supported Fe_3O_4 nanoparticles (Fe_3O_4/N-GAs; **Figure 16**) are known as efficient cathode catalysts for the oxygen reduction reaction (ORR) [50]. These hybrids were prepared via a combined hydrothermal self-assembly, freeze-drying, and thermal treatment process (**Figure 17**). The products showed an excellent electrocatalytic activity for the ORR in alkaline electrolytes, including a lower ring current, higher current density, higher electron transfer number (~4), lower H_2O_2 yield, and better durability.

3D graphene aerogels containing Fe_3O_4 nanoparticles (Fe_3O_4/GA (**Figure 18**), the lightest magnetic elastomer ever reported with density about 5.8 mg·cm^3) were hydrothermally prepared by self-assembly of graphene, simultaneously decorated with Fe_3O_4 nanoparticles [51]. They can be used to monitor the degree of compression/stretching of the material due to up to 52% reversible magnetic-field-induced strain and strain-dependent electrical resistance.

Among more complex aerogels, hydrophobic graphene aerogel/Fe_3O_4/polystyrene composites (having extremely low density of 0.005 g cm^{-3}, which corresponds to a volume porosity of 99.7%), with reticulated graphene structure, were solvothermally produced (**Figure 19**) [52]. Porous Fe_3O_4 nanoparticles were found to appear as partial substitutes for ethylenediamine-assisted cross-linking and interconnections between graphene plates. These composites were applied for crude oil remediation, allowing intake capacity as much as 40 times its own mass, after 10 water-oil separation cycles.

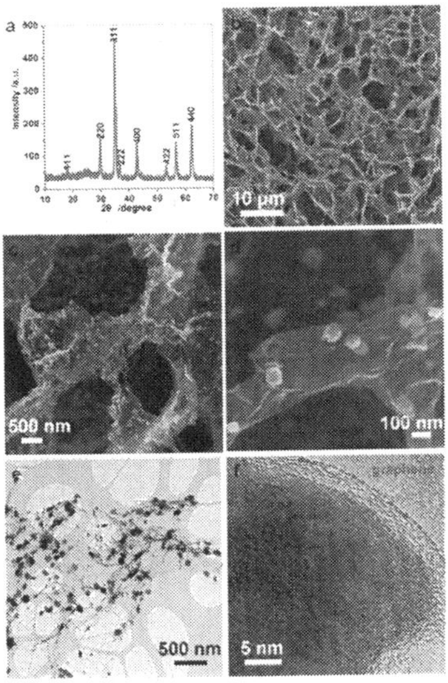

Figure 16. Structure and morphology of Fe_3O_4/N-GAs catalysts. (a) XRD pattern and (b–d) typical SEM images of Fe_3O_4/N-Gas, revealing the 3D macroporous structure and uniform distribution of Fe_3O_4 nanoparticles in the graphene aerogels. The red rings in (d) indicate Fe_3O_4 nanoparticles encapsulated in thin graphene layers. Representative (e) TEM and (f) HRTEM images of Fe_3O_4/N-GAs showing an Fe_3O_4 nanoparticle wrapped by graphene layers. Reproduced with permission from the *American Chemical Society*.

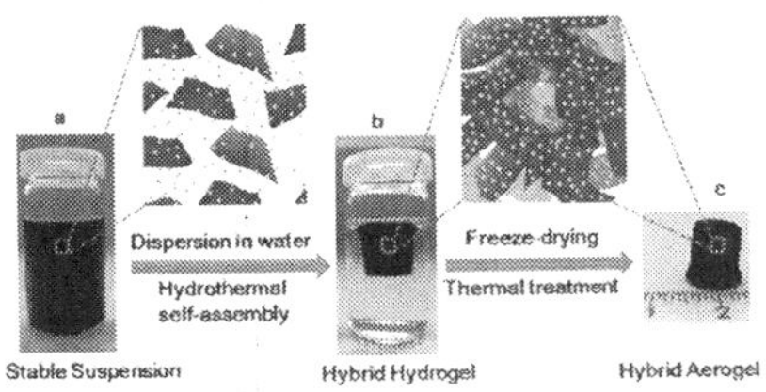

Figure 17. Fabrication process for the 3D Fe_3O_4/N-GAs catalyst. (a) Stable suspension of GO, iron ions, and polypyrrole (PPy) dispersed in a vial. (b) Fe- and PPy-supporting graphene hybrid hydrogel prepared by hydrothermal self-assembly and floating on water in a vial, and its ideal assembled model. (c) Monolithic Fe_3O_4/N-GAs hybrid aerogel obtained after freeze-drying and thermal treatment. Reproduced with permission from the *American Chemical Society*.

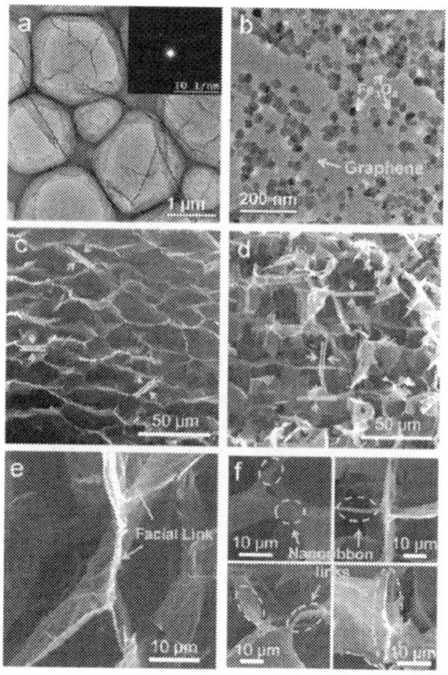

Figure 18. TEM and SEM images of the GA and Fe_3O_4/GA. (a) TEM image of GO and the diffraction pattern of a single flake (inset). (b) TEM image of Fe_3O_4 nanoparticle-decorated graphene sheets. (c, d) SEM images of microporous structures of GA and Fe_3O_4/GA. (e, f) SEM images of cross-linking patterns of GA and Fe_3O_4/GA. Reproduced with permission from the *American Chemical Society*.

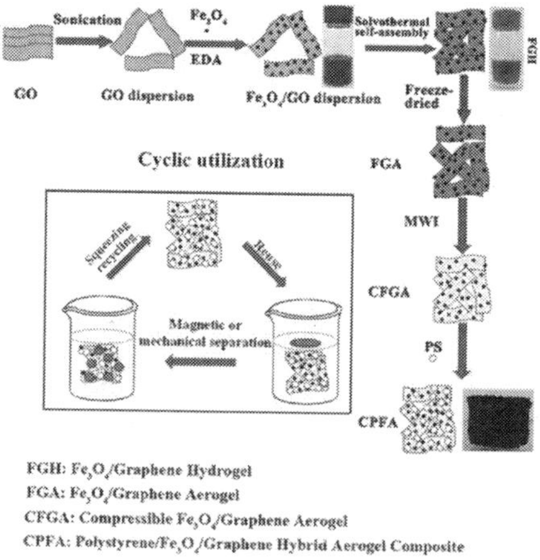

Figure 19. Schematic illustration of the fabrication process of the graphene aerogel/Fe_3O_4/polystyrene composites CPFA and its cyclic utilization for oil removal in water. Reproduced with permission from the *American Chemical Society*.

4.2. Bicomponent Fe_3O_4/G hybrids

Regarding the nonaerogel type of graphene-Fe_3O_4 hybrids, the hydrothermal [53] and solvothermal synthesis [54] have also been widely used although other methods are also frequently used, such as the atomic layer deposition, applied not only for Fe_3O_4/graphene but also for Ni/graphene composites [55]. Magnetic graphene foam with porous and hierarchical structures, on the basis of magnetite nanoparticles, was solvothermally obtained by gaseous reduction in a hydrothermal system and used for the adsorption of oil and organic solvents, thus serving for the cleanup of oil spills [56]. Fe_3O_4 nanoparticles on graphene foam possessed different morphologies, nanosheet arrays, or cubic structures, while controlling the reduction degree of graphene oxide under mild conditions. Distinct ratios "iron oxide/graphene" are important for several applications. Thus, G/Fe_3O_4 nanocomposites with different ratios of Fe_3O_4 to GO (mFe_3O_4: mG = 0.1:1, 0.2:1, 0.4:1, 0.6:1, 0.8:1 and 1:1) were prepared by solvothermal method and used for the removal of methylene blue dye from aqueous solutions [57]. The following morphologies were observed: uniform spherical homogeneously distributed Fe_3O_4 nanoparticles, with no agglomeration over the graphene sheets, and a uniform sheet-like shape of prepared graphene. Increasing the Fe_3O_4 nanoparticles on the surface of the graphene sheet was found to decrease the adsorption capacity, while the magnetization increased.

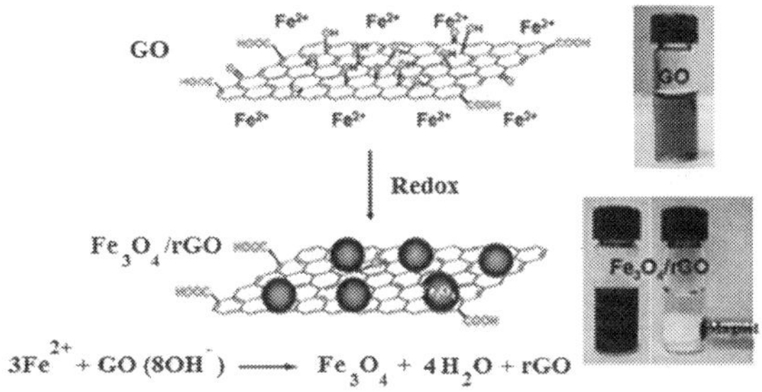

Figure 20. (a) A schematic representation of the preparation route of Fe_3O_4/rGO via a redox reaction between GO and Fe^{2+}. Photos show a water/NH_4OH (pH = 9) solution of Fe_3O_4/rGO (b) before and (c, left panel) after the redox reaction with Fe^{2+}, and (c, right panel) with an applied magnet. Reproduced with permission from the *Royal Society of Chemistry*.

Several methods, considered as "greener", are sometimes reported for iron oxide/graphene composites. Thus, the "green" oxidation of Fe^{2+} cations in $FeCl_2$ or $FeSO_4$ by graphene oxide led to an *in situ* deposition of Fe_3O_4 nanoparticles onto the self-reduced graphene oxide (rGO) sheets (**Figure 20**) [58]. Strongly supraparamagnetic with highly chemical reactivity, electrical conductivity, good solubility, and excellent processability G@Fe_3O_4 nanocomposites (with an average diameter of Fe_3O_4 nanoparticles of 1.2–6.3 nm; coverage density of Fe_3O_4 nanoparticles on graphene nanosheets of 5.3–57.9%) were prepared by a one-step "green" procedure (**Figure 21**) [59]. In addition, an approach for the deposition of iron oxide nanoparticles with *selective narrow size distribution* (0.5–7 and 1–3 μm), supported on different sizes of graphene oxide by coprecipitation, using Fe^{2+} and Fe^{3+} aqueous salt solutions and NH_3, is described in reaction (9) [60]. The reduction of mitochondrial activity using these materials is size dependent, but the chemical functionalization of GO and Fe_3O_4 is a way to enhance the biocompatibility, making the system independent of the size distribution of GO.

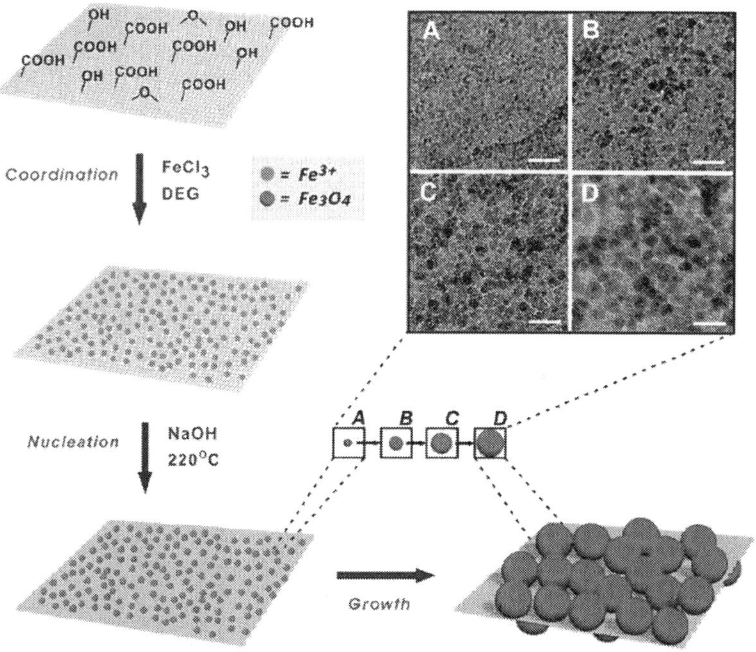

Figure 21. Schematic illustration of the formation of graphene@Fe_3O_4. The inset shows TEM images of graphene@ Fe_3O_4 at reaction times of 0 (A), 1 (B), 5 (C), and 60 min (D). Scale bars are 20 nm. Reproduced with permission from the *American Chemical Society*.

$(NH_4)_2Fe(SO_4)_2\cdot 6H_2O + 2NH_4Fe(SO_4)_2\cdot 12H_2O + 8NH_3\cdot H_2O \rightarrow Fe_3O_4\cdot 4H_2O + 6(NH_4)_2SO_4 + 14H_2O$ (9)

4.3. Multicomponent Fe_3O_4/G hybrids

More complex Fe_3O_4/G- and organic(inorganic)-containing systems are also common. Thus, coating a layer of mesoporous *silica* materials on each side of magnetic graphene, in the conditions of CTAB-assisted sol-gel process, with further calcination, led to obtaining magnetic graphene double-sided mesoporous nanocomposites (G/SiO_2) with high surface area (168 cm^2/g) and large pore volume (0.2 cm^3/g) (**Figure 22**) [61]. The formed materials were applied to size-selective and specific enrichment and identification of peptides (peptidomics analysis) in human urine samples, protein digest solutions, and standard peptide mixtures. Another example is G@mSiO_2-C18 materials (with a surface area of 315 $cm^2\cdot g^{-1}$ and a uniform pore size of 3.3 nm) with extended plate-like morphology, prepared by coating mesoporous silica layers onto graphene via surfactant-mediated cocondensation sol-gel process, and applied as magnetic solid-phase adsorbents to the selective enrichment of phthalates in water [62].

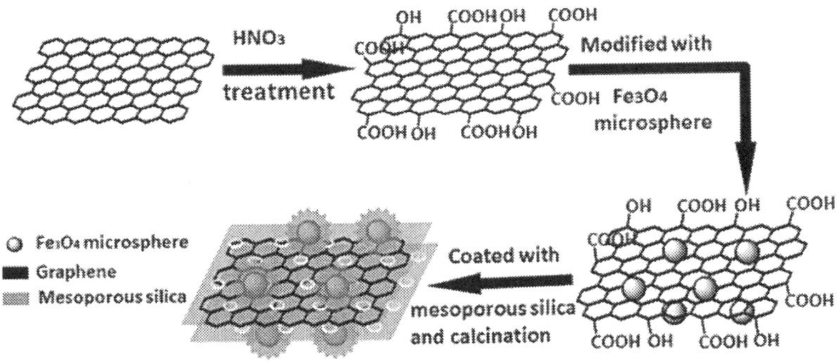

Figure 22. The workflow of synthesis of double-sided magnetic graphene/mSiO_2. Reproduced with permission from the *John Wiley & Sons*.

A few reports are dedicated to *chitosan-containing* magnetic materials with mainly biomedical applications. Thus, an iron oxide/graphene oxide/chitosan (Fe_3O_4/GO/CS) composite was obtained by a solution mixing-evaporation method [63]. Among other data, with the incorporation of 0.5 wt% Fe_3O_4 and 1 wt% GO, the tensile strength and Young's modulus of the composite significantly improved by about 28% and 74%, respectively, compared with chitosan. In addition, it was established from TGA that Fe_3O_4/GO/CS is less thermally stable than GO/CS composites, and graphite is more thermally stable than GO. In a related report [64], magnetic Fe_3O_4 nanoparticles were introduced into a water-dispersible and biocompatible chitosan-functionalized graphene, fabricated by a one-step ball milling of carboxylic chitosan and graphite. It could be an excellent catalyst for electrochemical biosensors, in particular for glucose detection, due to the presence of nitrogen (from chitosan) at the surface of graphene.

Also, superparamagnetic fluorescent Fe_3O_4/SiO_2/G-CdTe QD s/chitosan nanocomposites (Fe_3O_4/SiO_2/graphene-CdTe QDs/chitosan nanocomposites (FGQCs), with a spherical diameter of 467 nm; QDs: quantum dots) were prepared and studied for improving the drug loading content [65]. In addition, a magnetic composite bioadsorbent on the basis of magnetic chitosan and graphene oxide (MCGO) was prepared [66, 67]. Its adsorption capacity for methyl blue was found to be about 90% of the initial saturation adsorption capacity after being used four times. The adsorption of methyl blue on MCGO strongly depends on ionic strength and pH, showing an ion exchange mechanism.

In the case of *polymers*, integrated hybrid structures (stable for more than 1 year) consisting of exfoliated expanded graphite (EG flakes, both naked and functionalized with branched polyethylenimine (PEI)) and Fe_3O_4 nanocrystals were fabricated by an *ex situ* process by the integration of iron oxide nanoparticles, coated with meso-2,3-dimercaptosuccinic acid (DMSA) or poly(acrylic acid) (PAA), onto the exfoliated EG flakes under the support of *N*-hydroxysuccinnimide (NHS) and 1-ethyl-3-(3-dimethylaminopropyl)carbodiimide (EDC) [68]. Such materials can have antibacterial properties. For example, *in situ* growth of silver nanoparticles onto the polyethylenimine (PEI)-mediated magnetic reduced GO resulted in a bactericidal material, Ag@rGO-Fe_3O_4-PEI composite (**Figure 23**) [69]. The material provides a very high killing rate of 99.9% for *E. coli* bacteria under a 0.5 min near-infrared (NIR) laser irradiation.

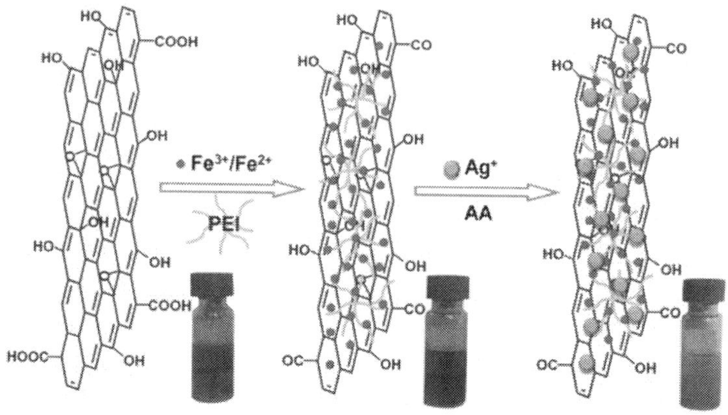

Figure 23. Schematic illustration of the preparation process of magnetic Ag@rGO-Fe_3O_4-PEI composites. Reproduced with permission from the *IOP Publishing*.

4.4. Applications of Fe_3O_4/G hybrids

Solvothermally prepared G/Fe_3O_4 (**Figure 24**) is able to effectively remove both bacteriophage and bacteria in water [70]. Indeed, it is capable of eliminating a wide range of pathogens including not only bacteriophages, but also various bacteria such as *Shigella, Salmonella, E. Faecium, E. coli, E. faecalis,* and *S. aureus,* with removal efficiencies up to 94%. Graphene oxide, particularly as magnetic Fe_3O_4/GO particles, was used as an adsorbent for wastewater treatment (**Figure 25**) [71]. A variety of other Fe_3O_4/G (or GO) hybrid applications are shown in **Table 1**.

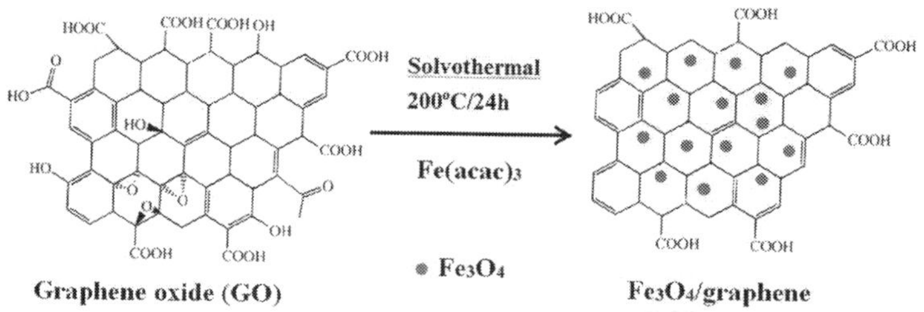

Figure 24. Synthesis route of G/Fe_3O_4. Reproduced with permission from the *American Chemical Society*.

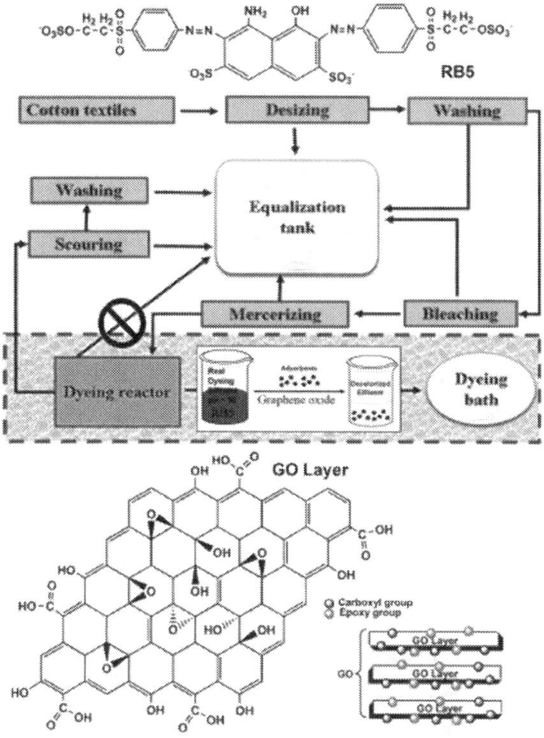

Figure 25. Illustrative scheme of the dyeing process and the proposed treatment of its effluents; RB5 dye (top) and functionalized GO adsorbent (bottom). Reproduced with permission from the *John Wiley & Sons*.

Composition	Description	Reference
Fe_3O_4/graphene aerogels	Removal of arsenic ions (up to 40.048 mg/g) from water	[72]
Fe_3O_4/graphene	Ultrahigh electrochemical sorption capacity for both inorganic arsenic species (arsenate As(V) and arsenite As(III)).	[73]
Magnetic graphene-oxide-supported-cyclodextrin	Heavy metal removal from wastewater. In particular, Cu(II) adsorption was found to be strongly pH-dependent and could be affected by background electrolytes, ionic strength, and citric acid.	[74]
Polyethylenimine(PEI)-modified magnetic mesoporous silica and graphene oxide (MMSP-GO)	These hierarchical composites are suitable for synergistic adsorptive simultaneous removal of heavy metal ions (Pb^{2+}, Cd^{2+}) and humic acid.	[75]
rGO/Fe_3O_4-based conductive and magnetic multifunctional films (membranes)	Can be produced and tested for its application in water desalination. Its potential uses also include catalysis, radiation shielding devices, biomedical fields, and supercapacitors.	[76]

Composition	Description	Reference
GO/Fe_3O_4	Super adsorbent to remove naphthalene, 1-naphthol, and 1-naphthylamine with different polarity. MWCNTs/FeO Fe_2O_3 were also produced in these synthesis as a byproduct, but graphene composite has the highest adsorption capacity among carbon-based nanomaterials. The adsorption capacity was shown to be naphthalene < 1-naphthol < 1-naphthylamine.	[77]
Magnetic graphene composite absorbent (Fe_3O_4@PDDA/GO@DNA) based on poly cationic core-shell Fe_3O_4@PDDA (poly(diallyldimethyl ammonium chloride) (PDDA)) and GO@DNA	Removal of trace levels of six types of polybrominated diphenyl ethers in water treatment. The nanomaterials can be reused at least 20 times for remediation purposes.	[78]
A porous graphene composite affinity material, containing graphene scaffolds, Fe_3O_4 nanoparticles, and fully covered porous titania nanostructures	Selective capture and convenient magnetic separation of target phosphopeptides, showing superior activity in comparison with commercial TiO_2 affinity materials.	[79]
3D graphene foam-supported Fe_3O_4	Use in lithium battery anodes.	[80]
The Fe_3O_4/G (graphene sheets) composite disks of μm dimensions	High-capacity lithium-ion battery anodes. The composites were prepared by electrostatic self-assembly between positively charged Fe_3O_4-APTMS [Fe_3O_4 grafted with (3-aminopropyl)trimethoxysilane (APTMS)] and negatively charged GO sheets and in an acidic solution (pH = 2) followed by *in situ* chemical reduction. An excellent rate capability as well as much enhanced cycling stability.	[81]

Table 1. Main applications of Fe_3O_4/G composites.

5. Ferrite-graphene and related composites

Simple or complex graphene or GO composites are known for a few ferrites, mainly those of cobalt and nickel, and usually used for environmental purposes or in batteries. $NiFe_2O_4$/G nanocomposites were prepared by mixing graphene and $NiFe_2O_4$ nanoparticles (obtained via a polyacrylamide gel method) into ethanol and further thermal drying at 60°C [82]. This composite exhibited significantly enhanced photocatalytic activity for the degradation of methylene blue in the conditions of irradiation of simulated sunlight, while $NiFe_2O_4$ nanoparticles were inert. NOH radicals were suggested as being the most active species causing dye degradation. $NiFe_2O_4$/GO nanohybrids based on $NiFe_2O_4$ and reduced GO consisted in nanosized $NiFe_2O_4$ crystals (5–10 nm in size) confined by few-layered rGO sheets [83]. This material seems a promising anode constituent for Li-ion batteries. The ternary nitrogen-doped graphene/nickel ferrite/polyaniline nanocomposite was hydrothermally prepared by a two-step approach process using urea, GO, $Ni(NO_3)_2 \cdot 6H_2O$, $Fe(NO_3)_3 \cdot 9H_2O$, and PANI as precursors, and applied in electrodes for supercapacitors [84]. It was confirmed that the introduction of N-heteroatom greatly improved the electrode specific capacitance.

In the case of cobalt, its graphene composites have been synthesized mainly by hydro/solvothermally although an ultrasonic method was also reported [85]. A $CoFe_2O_4$/G nanocomposite was hydrothermally prepared and was evaluated in the photocatalytic degradation of methylene blue dye under visible irradiation [86]. Its photocatalytic activity depends on the surface area and the structural and optical properties of samples, and behaves better than pure cobalt ferrite. The formation mechanism of hydrothermally synthesized $CoFe_2O_4$-FGS (FGS: functionalized graphene sheets) is described in detail by Li et al. [87]. In addition, porous $CoFe_2O_4$/rGO nanoclusters with different concentrations of graphene were solvothermally prepared (**Figure 26**) [88] and their electrochemical properties were evaluated using polyvinylidene fluoride and Na-alginate as binder materials to improve the anode performance of Li-ion batteries. The resulting $CoFe_2O_4$/rGO (20%) nanocomposites with sodium alginate binders exhibited promising electrochemical performance, such as excellent recycling behavior, good rate capability, and high reversible capacity.

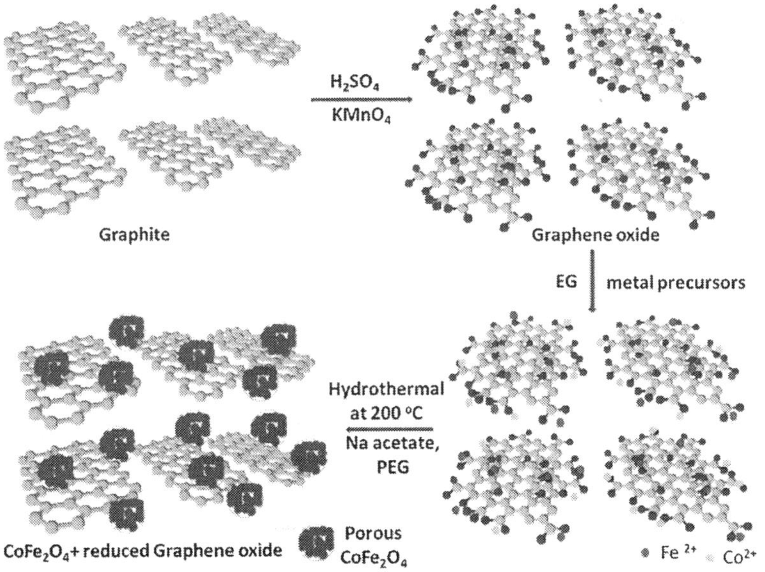

Figure 26. Synthesis of the $CoFe_2O_4$/rGO composite. Reproduced with permission from the *Royal Society of Chemistry*.

Other graphene- (or GO)-ferrite hybrids are rare (for example, zinc-ferrite-reduced graphene oxide [89]), and their synthesis, properties, and applications are similar to those of nickel and cobalt. Thus, in a solvothermally fabricated uniform copper ferrite nanoparticle-attached graphene nanosheet ($CuFe_2O_4$/G), $CuFe_2O_4$ nanoparticles (with a diameter of about 100 nm) were shown to be densely and compactly deposited on graphene nanosheets [90]. This nanomaterial exhibited properties of a high-performance supercapacitor material. A related hydrothermal route was used for the formation of $CuFe_2O_4$ nanoparticles into the GO sheets, and the obtained material, with GO content varying from 10 to 30 wt%, was used as stable and good reproducible enzyme-free glucose sensor [91]. $MnFe_2O_4$ nanoparticle (MFNP)-decorated GO nanocomposites (MGONCs) were prepared through a miniemulsion and solvent evaporation process [92]. The loading of magnetic nanocrystals depends on the ratio of GO/magnetic nanoparticles. In a related report [93], $MnFe_2O_4$ nanoparticles were deposited on GO by thermal decomposition of manganese(II) acetylacetonate and iron(III) acetylacetonate precursors in triethylene glycol. The resulting GO/$MnFe_2O_4$ nanohybrids showed imperceptible *in vivo* toxicity, negligible hemolytic activity, and very low cytotoxicity, being effective T2 contrast agents.

In addition, barium hexaferrite nanoparticles, fabricated [94] via the citrate sol-gel combustion method, in a reaction medium consisting of various forms of graphene nanosheets (such as reduced graphite oxide, expanded graphite oxide, and expanded graphite) were used to prepare nanocomposites as microwave absorbing material for radars [95]. Finally, $BiFeO_3$ nanoparticles that are important materials in photocatalysis [96] were prepared via a polyacrylamide gel route and transformed into a $BiFeO_3$/G nanocomposite, by mixing with graphene in ethanol and further thermal drying at 60°C [97]. Methyl orange (MO) was found to be efficiently degraded by this material under simulated sunlight irradiation. This photocatalytic performance can be mainly ascribed to the efficient transfer of photogenerated electrons from $BiFeO_3$ to graphene (**Figure 27**). A similar reduced graphene rGO/$BiFeO_3$ nanocomposite, prepared by self-assembly, was also reported [98].

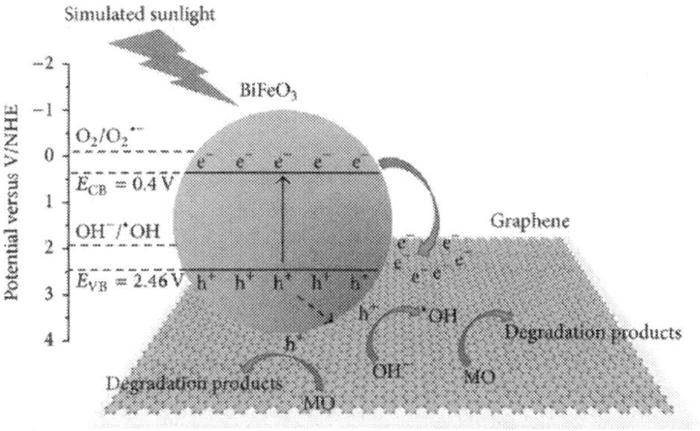

Figure 27. Schematic illustration of the photocatalytic mechanism of $BiFeO_3$-graphene nanocomposite toward the degradation of MO. Reproduced with permission from the *Hindawi Publishing Corporation.*

An interesting example of a magnetic molecule and its composite with graphene is known for lanthanides. A device made of a graphene nanoconstruction, decorated with $TbPc_2$ magnetic molecules (Pc: phthalocyananine), was able to electrically detect the magnetization reversal of the molecules in proximity with graphene (**Figure 28**) [99, 100].

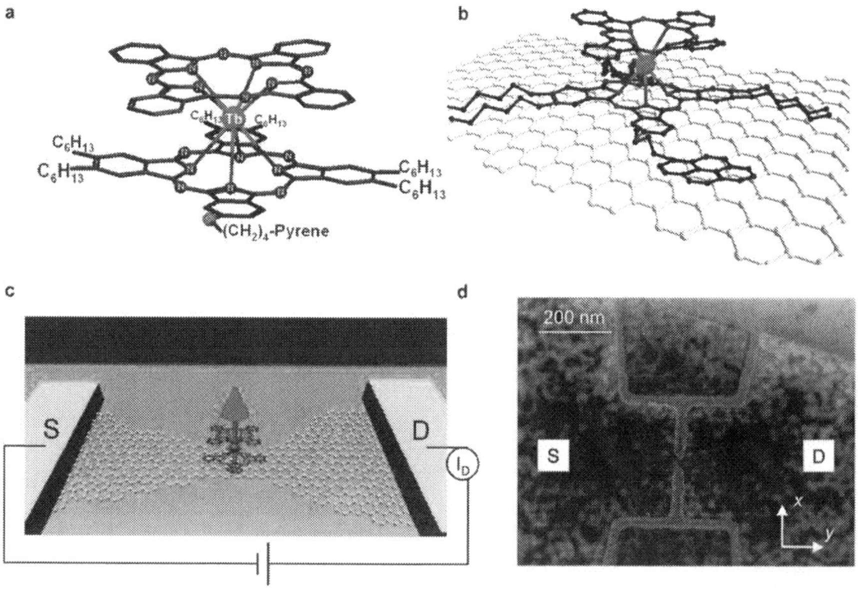

Figure 28. (a) Schematic representation of the $TbPc_2$ single-molecule magnet (SMM). (b, c) Schematic view of the device, showing in (b) the molecule attached to graphene and in (c) the nanoconstriction contacted by source (S) and drain (D) electrodes. The magnetic moments of the $TbPc_2$ SMMs (hexyl and pyrenyl groups here omitted for clarity) on top of the constriction add another degree of freedom to tune the conductivity of the device. (d) False-colored SEM image of the device presented in the text. SiO_2 substrate and etched graphene are colored in purple, while graphene conductive regions are colored in green. Source and drain electrodes are indicated. Reproduced with permission from the *American Chemical Society*.

6. Conclusions and outlook

The field of graphene and graphene oxide composites containing magnetic components can be considered as an important subject of the modern nanotechnology. Similarly to the filled or decorated carbon nanotubes, for which endohedral and exohedral functionalization with magnetic nanoparticles is well known, the graphene and graphene oxide magnetic nanocomposites can also be obtained via distinct methods and have a variety of applications. Graphene magnetic nanocomposites are currently known for those on the basis of elemental metals (Fe, Co, and Ni), magnetic nanoclusters, iron oxides (Fe_2O_3, Fe_3O_4) in distinct morphological forms, ferrites MFe_2O_4, 3D graphene aerogels@hierarchical Fe_3O_4 nanoclusters, single-molecule magnets such as $TbPc_2$ (Pc: phthalocyanine), other organometallic-containing composites (benzene-metal-graphene), as well as polycomponent nanocomposites such as $Ag/Fe_3O_4/G$ (G: graphene), $Fe_3O_4/CdS/G$, or $FePc/Fe_3O_4/GO$ (GO :graphene oxide), etc.

Magnetic graphene and graphene oxide hybrids can be prepared by a variety of methods. Hydrothermal and solvothermal techniques are the most common. Other routes are also available, such as sol-gel autocombustion, sonoelectrochemical polymerization, thermal expansion and thermal reduction, microwave-assisted technique, as well as covalent bonding chemical methods. The formed composites have or might have several current and/or potential applications, such as devices (supercapacitors or material for anodes of lithium batteries), construction (graphene oxide-ferrofluid-cement nanocomposites), analytical, sensor, and biosensor applications (graphene-based magnetic solid-phase extraction for the determination of five chloroacetanilide herbicides (alachlor, acetochlor, metolachlor, butachlor, and pretilachlor), nitrite sensor, composites with peroxidase-like activity for colorimetric detection of glucose, etc.), environmental remediation (degradation and removal of water pollutants, such as chlorophenols, adsorbent for pesticide extraction, As and heavy metals (U^{VI})), compounds with antibacterial properties, and, of course, as magnetic nanocatalysts or nanophotocatalysts which can be simply magnetically removed from reaction media [101, 102].

As described above, magnetic graphene and graphene oxide composites can have applications, in particular, in the removal, decoloration, sequestration and adsorption of dyes [103], antibiotics and other organic contaminants, heavy metal separation (Hg, Cd, Cr, Cu, and As), and other environmental remediation aspects [104]. Other areas include biological imaging and further bioapplications [105], magnetic separation, electromagnetic materials and coatings, conducting polymer nanocomposites, aligned substrate for nanodevices, and spintronic devices with perpendicular magnetic anisotropy. 3D graphene aerogels can have applications in electrode materials, supercapacitors, oil absorption, and gas sensors due to their electrical conductivity, mechanical strength, and high porosity.

Author details

Oxana Vasilievna Kharissova[1], Beatriz Ortega García[1], Boris Ildusovich Kharisov[2*] and Ubaldo Ortiz Méndez[3]

*Address all correspondence to: bkhariss@hotmail.com

1 Department of Physico-Mathematics, Autonomous University of Nuevo León, Monterrey, Mexico

2 Department of Chemistry, Autonomous University of Nuevo León, Monterrey, Mexico

3 FIME (Department of Mechanical & Electrical Engineering), Autonomous University of Nuevo León, Monterrey, Mexico

References

[1] Liu, X.; Wang, C.-Z.; Hupalo, M.; Lin, H.-Q.; Ho, K.M.; Tringides, M.C. Metals on graphene: interactions, growth morphology, and thermal stability. Crystals, 2013, 3, 79–111.

[2] Lim, H.N.; Huang, N.M.; Chia, C.H.; Harrison, I.I. Inorganic nanostructures decorated graphene. In: Advanced Topics on Crystal Growth (Ed. Sukarno Olavo Ferreira), Intech, Rijeka, Croatia, 2013.

[3] Dahal, A.; Batzill, M. Graphene–nickel interfaces: a review. Nanoscale, 2014, 6, 2548–2562.

[4] Dhand, V.; Rhee, K.Y.; Kim, H.J.; Jung, D.H. A comprehensive review of graphene nanocomposites: research status and trends, J. Nanomater., 2013, 763953, 14 pp.

[5] Masotti, A.; Caporali, A. Preparation of magnetic carbon nanotubes (Mag-CNTs) for biomedical and biotechnological applications. Int. J. Mol. Sci., 2013, 14, 24619–24642.

[6] Hu, F.M.; Gubernatis, J.E.; Lin, H.-Q.; Li, Y.-C.; Nieminen, R.M. Behavior of a magnetic impurity in graphene in the presence of a vacancy. Phys. Rev. B, 2012, 85(11), 115442/1-5.

[7] Hu, F.M.; Ma, T.; Lin, H.-Q.; Gubernatis, J.E. Magnetic impurities in graphene. Phys. Rev. B, 2011, 84, 075414.

[8] Voloshina, E.; Dedkov, Yu. Electronic and magnetic properties of the graphene-ferromagnet interfaces: theory vs. experiment. In: Physics and Applications of Graphene—Experiments (Ed. Mikhailov, S.), Intech, Rijeka, Croatia, 2011.

[9] Le, H.M.; Hirao, H.; Kawazoe, Y.; Nguyen-Manh, D.J. Nanostructures of C_{60}-metal-graphene (metal = Ti, Cr, Mn, Fe, or Ni): a spin-polarized density functional theory study. Phys. Chem. C, 2014, 118, 21057–21065.

[10] Raji, A.T.; Lombardi, E.B. Stability, magnetic and electronic properties of cobalt–vacancy defect pairs in graphene: A first-principles study. Physica B, 2015, 464, 28–37.

[11] Li, B.; Xu, D.; Zhao, J.; Zeng, H. First principles study of electronic and magnetic properties of Co-doped armchair graphene nanoribbons. J. Nanomater., 2015, 538180, 9 pp.

[12] Donati, F.; Dubout, Q.; Autes, G.; Patthey, F.; Calleja, F.; Gambardella, P.; Yazyev, O.V.; Brune, H. Magnetic moment and anisotropy of individual Co atoms on graphene. Phys. Rev. Lett., 2013, 236801, 5 pp.

[13] Vita, H.; Bottcher, St.; Leicht, P.; Horn, K.; Shick, A.B.; Maca, F. Electronic structure and magnetic properties of cobalt intercalated in graphene on Ir(111). Phys. Rev. B, 2014, 90, 165432.

[14] Decker, R.; Brede, J.; Atodiresei, N.; Caciuc, V.; Blugel, S.; Wiesendanger, R. Atomic-scale magnetism of cobalt-intercalated graphene. Phys. Rev. B, 2013, 87, 041403(R).

[15] Rougemaille, N.; N'Diaye, A.T.; Coraux, J.; Vo-Van, C.; Fruchart, O.; Schmid, A.K. Perpendicular magnetic anisotropy of cobalt films intercalated under graphene. Appl. Phys. Lett., 2012, 101, 142403.

[16] Vermisoglou, E.C.; Devlin, E.; Giannakopoulou, T.; Romanos, G.; Boukos, N.; Psycharis, V.; Lei, C.; Lekakou, C.; Petridis, D.; Trapalis, C. Reduced graphene oxide/iron carbide nanocomposites for magnetic and supercapacitor applications. J. Alloys Compounds, 2014, 590, 102–109.

[17] Afshar, M.; Doosti, H. Magnetic properties of cobalt single layer added on graphene: a density functional theory study. Mod. Phys. Lett. B, 2015, 29(2), 1450262, 9 pp.

[18] Zhidkov, I.S.; Skorikov, N.A.; Korolev, A.V.; Kukharenko, A.I.; Kurmaev, E.Z.; Fedorov, V.E.; Cholakh, S.O. Electronic structure and magnetic properties of graphene/Co composites. Carbon, 2015, 91, 298–303.

[19] Lin, K.-Y.A.; Hsu, F.-K.; Lee, W.-D. Magnetic cobalt–graphene nanocomposite derived from self-assembly of MOFs with graphene oxide as an activator for peroxymonosulfate. J. Mater. Chem. A, 2015, 3, 9480–9490.

[20] Zhang, L.; Huang, Y.; Zhang, Y.; Ma, Y.; Chen, Y. Sol–gel autocombustion synthesis of graphene/cobalt magnetic nanocomposites. J. Nanosci. Nanotechnol., 2013, 13, 1129–1131.

[21] Yang, S.; Cui, G.; Pang, S.; Cao, Q.; Kolb, U.; Feng, X.; Maier, J.; Mullen, K. Fabrication of cobalt and cobalt oxide/graphene composites: towards high-performance anode materials for lithium ion batteries. ChemSusChem, 2010, 3, 236–239.

[22] Yao, Y.; Xu, C., Qin, J., Wei, F.; Rao, M.; Wang, S. Synthesis of magnetic cobalt nanoparticles anchored on graphene nanosheets and catalytic decomposition of Orange II. Ind. Eng. Chem. Res. 2013, 52, 17341–17350.

[23] Umair, A.; Raza, H. Controlled synthesis of bilayer graphene on nickel. Nanoscale Res. Lett., 2012, 7, 437.

[24] Dlubak, B.; Martin, M.-B.; Weatherup, R.S., Yang, H.; Deranlot, C.; Blume, R.; Schloegl, R.; Fert, A.; Anane, A.; Hofmann, S.; Seneor, P.; Robertson, J. Graphene-passivated nickel as an oxidation-resistant electrode for spintronics. ACS Nano, 2012, 6(12), 10930–10934.

[25] Sicot, M.; Bouvron, S.; Zander, O.; Rudiger, U.; Dedkov, Vu.S.; Fonin, M. Nucleation and growth of nickel nanoclusters on graphene Moire on Rh(111). Appl. Phys. Lett., 2010, 96, 093115.

[26] Mural, P.K.S.; Pawar, S.P.; Jayanthi, S.; Madras, G.; Sood, A.K.; Bose, S. Engineering nanostructures by decorating magnetic nanoparticles onto graphene oxide sheets to shield electromagnetic radiations. ACS Appl. Mater. Interf., 2015, 7, 16266–16278.

[27] Qu, W.; Zhang, L.; Chen, G. Magnetic loading of graphene–nickel nanoparticle hybrid for electrochemical sensing of carbohydrates. Biosens. Bioelectron., 2013, 42, 430–433.

[28] Fang, J.; Zha, W.; Kang, M.; Lu, S.; Cui, L.; Li, S. Microwave absorption response of nickel/graphene nanocomposites prepared by electrodeposition. J. Mater. Sci., 2013, 48, 8060–8067.

[29] Aravind, S.S.J.; Eswaraiah, V.; Ramaprabhu, S. Facile and simultaneous production of metal/metal oxide dispersed graphene nano composites by solar exfoliation. J. Mater. Chem. 2011, 21(43), 17094–17097.

[30] Lee, S.-H.; Sridhar, V.; Jung, J.-H.; Karthikeyan, K.; Lee, Y.S.; Mukherjee, R.; Koratkar, N.; Oh, I.-K. Graphene nanotube iron hierarchical nanostructure as lithium ion battery anode. ACS Nano, 2013, 785, 4242–4251.

[31] Takahashi, K.; Wang, Y.; Chiba, S.; Nakagawa, Y.; Isobe, S., Ohnuki, S. Low temperature hydrogenation of iron nanoparticles on graphene. Sci. Rep., 2014, 4, 4598.

[32] Kumar, H.; Shukla, A.K. Fabrication Fe/Fe_3O_4/graphene nanocomposite electrode material for rechargeable Ni/Fe batteries in hybrid electric vehicles. Int. Lett. Chem. Phys. Astron., 2013, 2013, 15–25.

[33] Jabeen, H.; Chandra, V., Jung, S.; Woo Lee, J.; Kim, K.S.; Bin Kim, S. Enhanced Cr(VI) removal using iron nanoparticle decorated graphene. Nanoscale, 2011, 3, 3583–3585.

[34] Guo, J.; Wang, R.; Weei Tjiu, W.; Pan, J.; Liu, T. Synthesis of Fe nanoparticles@graphene composites for environmental applications. J. Hazardous Mater., 2012, 225–226, 63–73.

[35] Li, Y.; Zhou, W.; Wang, H.; Xie, L.; Liang, Y.; Wei, F.; Idrobo, J.-C.; Pennycook, S.J.; Dai, H. An oxygen reduction electrocatalyst based on carbon nanotube–graphene complexes. Nat. Nanotechn., 2012, 7, 394–400.

[36] Parvez, K.; Yang, S., Hernandez, Y.; Winter, A.; Turchanin, A.; Feng, X.; Mullen, K. Nitrogen-doped graphene and its iron-based composite as efficient electrocatalysts for oxygen reduction reaction. ACS Nano, 2012, 6(11), 9541–9550.

[37] Zhu, J.; Sadu, R.; Wei, S.; Chen, D.H.; Haldolaarachchige, N.; Luo, Z.; Gomes, J.A.; Young, D.P.; Guo, Z. Magnetic graphene nanoplatelet composites toward arsenic removal. ECS J. Solid State Sci. Technol., 2012, 1(1), M1–M5.

[38] Zhu, J.; Wei, S.; Gu, H.; Rapole, S.B.; Wang, Q.; Luo, Z.; Haldolaarachchige, N.; Young, D.P.; Guo, Z. One-pot synthesis of magnetic graphene nanocomposites decorated with core@double-shell nanoparticles for fast chromium removal. Environ. Sci. Technol. 2012, 46, 977–985.

[39] Zhang, X.; Alloul, O.; He, Q.; Zhu, J.; Joseph Verde, M.; Li, Y.; Wei, S.; Guo, Z. Strengthened magnetic epoxy nanocomposites with protruding nanoparticles on the graphene nanosheets. Polymer, 2013, 54, 3594–3604.

[40] Wang, Y.; He, Q.; Qu, H.; Zhang, X.; Guo, J.; Zhu, J.; Zhao, G.; Colorado, H.A.; Yu, J. et al. Magnetic graphene oxide nanocomposites: nanoparticles growth mechanism and property analysis. J. Mater. Chem. C, 2014, 2, 9478–9488.

[41] McCoy, T.M.; Brown, P.; Eastoe, J.; Tabor, R.F. Noncovalent magnetic control and reversible recovery of graphene oxide using iron oxide and magnetic surfactants. ACS Appl. Mater. Interf., 2015, 7, 2124–2133.

[42] Sun, Z.; Madej, E.; Wiktor, C. et al. One-pot synthesis of carbon-coated nanostructured iron oxide on few-layer graphene for lithium-ion batteries. Chem. Eur. J., 2015, 21, 16154–16161.

[43] Jang, B.; B. Chae, O.; Park, S.-K.; Ha, J.; Oh, S.-M.; Na, H.B.; Piao, Y. Solventless synthesis of an iron-oxide/graphene nanocomposite and its application as an anode in high-rate Li-ion batteries. J. Mater. Chem. A, 2013, 1, 15442–15446.

[44] Tae Kim, I.; Magasinski, A.; Jacob, K.; Yushin, G.; Tannenbaum, R. Synthesis and electrochemical performance of reduced graphene oxide/maghemite composite anode for lithium ion batteries. Carbon, 2013, 52, 56–64.

[45] Zhu, X.; Zhu, Y.; Murali, S.; Stoller, M.D.; Ruoff, R.S. Nanostructured reduced graphene oxide/Fe_2O_3 composite as a high-performance anode material for lithium ion batteries. ACS Nano, 2011, 5(4), 3333–3338.

[46] Eeu, Y.C.; Lim, H.N.; Lim, Y.S.; Zakarya, S.A.; Huang, N.M. Electrodeposition of polypyrrole/reduced graphene oxide/iron oxide nanocomposite as supercapacitor electrode material. J. Nanomater., 2013, 2013, 653890, 6 pp.

[47] Sinh, A.; Jan, N.R. Separation of microcystin-LR by cyclodextrin-functionalized magnetic composite of colloidal graphene and porous silica. ACS Appl. Mater. Interf., 2015, 7, 9911–9919.

[48] Jagadeesh, R.V.; Natte, K.; Junge, H.; Beller, M. Nitrogen-doped graphene-activated iron-oxide-based nanocatalysts for selective transfer hydrogenation of nitroarenes. ACS Catal., 2015, 5, 1526–1529.

[49] Jung, S.M.; Jung, H.Y.; Dresselhaus, M.S.; Jung, J.J.; Kong, J. A facile route for 3D aerogels from nanostructured 1D and 2D materials. Sci. Rep., 2012, 2, 849.

[50] Wu, Z.-S.; Yang, S.; Sun, Y.; Parvez, K.; Feng, X.; Mu®llen, K. 3D nitrogen-doped graphene aerogel-supported Fe_3O_4 nanoparticles as efficient electrocatalysts for the oxygen reduction reaction. J. Am. Chem. Soc., 2012, 134, 9082–9085.

[51] Xu, X.; Li, H.; Zhang, Q.; Hu, H.; Zhao, Z.; Li, J.; Li, J.; Qiao, Yu.; Gogotsi, Yu. Self-sensing, ultralight, and conductive 3D graphene/iron oxide aerogel elastomer deformable in a magnetic field. ACS Nano, 2015, 9(4), 3969–3977.

[52] Zhou, S.; Jiang, W.; Wang, T.; Lu, Y. Highly hydrophobic, compressible, and magnetic polystyrene/Fe_3O_4/graphene aerogel composite for oil–water separation. Ind. Eng. Chem. Res., 2015, 54, 5460–5467.

[53] Ren, L.; Huang, S.; Fan, W.; Liu, T. One-step preparation of hierarchical superparamagnetic iron oxide/graphene composites via hydrothermal method. Appl. Surf. Sci., 2011, 258, 1132–1138.

[54] Fu, C.; Zhao, G.; Zhang, H.; Li, S. A facile route to controllable synthesis of Fe_3O_4/graphene composites and their application in lithium-ion batteries. Int. J. Electrochem. Sci., 2014, 9, 46–60.

[55] Wang, G.; Gao, Z.; Wan, G.; Lin, S.; Yang, P.; Qin, Y. Supported high-density magnetic nanoparticles on graphene by atomic layer deposition used as efficient synergistic microwave absorbers. Nanoresearch, 2014, 7(5), 704–716.

[56] Yang, S.; Chen, L.; Mu, L.; Ma, P.C. Magnetic graphene foam for efficient adsorption of oil and organic solvents. J. Colloid Interf. Sci., 2014, 430, 337–344.

[57] Farghali, M.A.; Salah El-Din, T.A.; Al-Enizi, A.M.; El Bahnasawy, R.M. Graphene/magnetite nanocomposite for potential environmental application. Int. J. Electrochem. Sci., 2015, 10, 529–537.

[58] Xue, Y.; Chen, H.; Yu, D. et al. Oxidizing metal ions with graphene oxide: the in situ formation of magnetic nanoparticles on self-reduced graphene sheets for multifunctional applications. Chem. Commun., 2011, 47, 11689–11691.

[59] He, H.; Gao, C. Supraparamagnetic, conductive, and processable multifunctional graphene nanosheets coated with high-density Fe_3O_4 nanoparticles. Appl. Mater. Interf., 2010, 2(11), 3201–3210.

[60] Urbas, K.; Aleksandrzak, M.; Jedrzejczak, M.; Jedrzejczak, M.; Rakoczy, R.; Chen, X.; Mijowska, E. Chemical and magnetic functionalization of graphene oxide as a route to enhance its biocompatibility. Nanoscale Res. Lett., 2014, 9, 656.

[61] Yin, P.; Sun, N.; Deng, C.; Li, Y.; Zhang, X.; Yang, P. Facile preparation of magnetic graphene double-sided mesoporous composites for the selective enrichment and analysis of endogenous peptides. Proteomics, 2013, 13, 2243–2250.

[62] Huang, D.; Wang, X.; Deng, C.; Song, G.; Cheng, H.; Zhang, X. Facile preparation of raisin-bread sandwich-structured magneticgraphene/mesoporous silica composites with C18-modified pore-walls for efficient enrichment of phthalates in environmental water. J. Chromatogr. A, 2014, 1325, 65–71.

[63] Yadav, M.; Rhee, K.Y.; Park, J.S.; Hui, D. Mechanical properties of Fe_3O_4/GO/chitosan composites. Composites: Part B, 2014, 66, 89–96.

[64] Zhang, W.; Li, X.; Zou, R.; Wu, H.; Shi, H.; Yu, S.; Liu, Y. Multifunctional glucose biosensors from Fe_3O_4 nanoparticles modified chitosan/graphene nanocomposites. Sci. Rep., 2015, 5, 11129.

[65] Ou, Y.; Wang, F.; Huang, Y.; Li, D.; Jiang, Y.; Qin, Q.-H.; Stachurski, Z.H.; Tricoli, A.; Zhang, T. Fabrication and cyto-compatibility of Fe_3O_4/SiO_2/graphene–CdTeQDs/CS nanocomposites for drug delivery. Colloids Surf. B: Biointerf., 2014, 117, 466–472.

[66] Fan, L.; Luo, C.; Li, X.; Lu, F.; Qiu, H.; Sun, M. Fabrication of novel magnetic chitosan grafted with graphene oxide to enhance adsorption properties for methyl blue. J. Hazardous Mater., 2012, 215–216, 272–279.

[67] Fan, L.; Luo, C.; Sun, M.; Li, X.; Lu, F.; Qiu, H. Preparation of novel magnetic chitosan/graphene oxide composite as effective adsorbents toward methylene blue. Bioresource Technol., 2012, 114, 703–706.

[68] Jovici, N.; Calatayud, M.P.; Sanz, B.; Montone, A.; Goya, G.F. Ex situ integration of iron oxide nanoparticles onto exfoliated expanded graphite flakes in aqueous suspension. J. Serb. Chem. Soc., 2014, 79(9), 1155–1167.

[69] Wang, N.; Hu, B.; Chen, M.-L.; Wang, J.-H. Polyethylenimine mediated silver nanoparticle-decorated magnetic graphene as a promising photothermal antibacterial agent. Nanotechnology, 2015, 26, 195703, 8 pp.

[70] Zhan, S.; Zhu, D.; Ma, S.; Yu, W.; Jia, Y.; Li, Y.; Yu, H.; Shen, Z. Highly efficient removal of pathogenic bacteria with magnetic graphene composite. ACS Appl. Mater. Interf., 2015, 7, 4290–4298.

[71] Kyzas, G.Z.; Deliyanni, E.A.; Matis, K.A. Graphene oxide and its application as an adsorbent for wastewater treatment. J. Chem. Technol. Biotechnol., 2014, 89, 196–205.

[72] Ye, Y.; Yin, D.; Wang, B.; Zhang, Q. Synthesis of three-dimensional Fe_3O_4/graphene aerogels for the removal of arsenic ions from water. J. Nanomater., 2015, 864864, 6 pp.

[73] Mishra, A.K.; Ramaprabhu, S. Ultrahigh arsenic sorption using iron oxide-graphene nanocomposite supercapacitor assembly. J. Appl. Phys., 2012, 112, 104315.

[74] Hu, X.-j.; Liu, Y.-g.; Wang, H.; Zeng, G.-m.; Hua, X.; Guo, Y.-m.; Li, T.-t.; Chen, A.-w.; Jiang, L.-h.; Guo, F.-y. Adsorption of copper by magnetic grapheneoxide-supported -cyclodextrin: Effects of pH, ionicstrength, background electrolytes, and citric acid. Chem. Chem. Eng. Res. Design, 2015, 93, 675–683.

[75] Wang, Y.; Liang, S.; Chen, B.; Guo, F.; Yu, S.; Tang, Y. Synergistic removal of Pb(II), Cd(II) and humic acid by Fe_3O_4@mesoporous silica-graphene oxide composites. PLOS ONE, 2013, 8(6), e65634.

[76] Narayanan, T.N.; Liu, Z.; Lakshmy, P.R.; Gao, W.; Nagaoka, Y.; Sakthi Kumar, D.; Lou, J.; Vajtai, R.; Ajayan, P.M. Synthesis of reduced graphene oxide–Fe_3O_4 multifunctional

freestanding membranes and their temperature dependent electronic transport properties. Carbon, 2012, 50, 1338–1345.

[77] Yang, X.; Li, J.; Wen, T.; Ren, X., Huang, Y.; Wang, X. Adsorption of naphthalene and its derivatives on magnetic graphene composites and the mechanism investigation. Colloids Surf. A: Physicochem. Eng. Aspects, 2013, 422, 118–125.

[78] Gan, N.; Zhang, J.; Lin, S.; Long, N.; Li, T.; Cao, Y. A novel magnetic graphene oxide composite absorbent for removing trace residues of polybrominated diphenyl ethers in water. Materials 2014, 7, 6028–6044.

[79] Cheng, G.; Yu, X.; Zhou, M.D.; Zheng, S.-Y. Preparation of magnetic graphene composites with hierarchical structure for selective capture of phosphopeptides. J. Mater. Chem. B, 2014, 2, 4711–4719.

[80] Luo, J.; Liu, J.; Zeng, Z. et al. Three-dimensional graphene foam supported Fe_3O_4 lithium battery anodes with long cycle life and high rate capability. Nano Lett., 2013, 13, 6136–6143.

[81] Yoon, T.; Kim, J.; Kim, J.; Kyoo Lee, J. Electrostatic self-assembly of Fe_3O_4 nanoparticles on graphene oxides for high capacity lithium-ion battery anodes. Energies, 2013, 6, 4830–4840.

[82] Zhao, D.F.; Yang, H., Li, R.S.; Ma, J.Y.; Feng, W.J. Fabrication of nickel ferrite–graphene nanocomposites and their photocatalytic properties. Mater. Res. Innovations, 2014, 18(7), 519–523.

[83] Zhu, P.; Liu, S.; Xie, J.; Zhang, S.; Cao, G.; Zhao, X. Facile synthesis of $NiFe_2O_4$/reduced graphene oxide hybrid with enhanced electrochemical lithium storage performance. J. Mater. Sci. Technol., 2014, 30(11), 1078–1083.

[84] Wang, W.; Hao, Q.; Lei, W.; Xia, X.; Wang, X. Ternary nitrogen-doped graphene/nickel ferrite/polyaniline nanocomposites for high-performance supercapacitors. J. Power Sources, 2014, 269, 250–259.

[85] Xiao, Y.; Li, X.; Zai, J.; Wang, K.; Gong, Y.; Li, B.; Han, Q., Qian, X. $CoFe_2O_4$-graphene nanocomposites synthesized through an ultrasonic method with enhanced performances as anode materials for Li-ion batteries. Nano-Micro Lett., 2014, 6(4), 307–315.

[86] Suwanchawalit, C.; Somjit V. Hydrothermal synthesis of magnetic $CoFe_2O_4$-graphene nanocomposite with enhanced photocatalytic performance. Digest J. Nanomater. Biostruct, 2015, 10(3), 769–777.

[87] Li, N.; Zheng, M.; Chang, X.; Ji, G.; Lu, H.; Xue, L.; Pan, L.; Cao, J. Preparation of magnetic $CoFe_2O_4$-functionalized graphene sheets via a facile hydrothermal method and their adsorption properties. J. Solid State Chem., 2011, 184, 953–958.

[88] Ramesh Kumar, P.; Kollu, P.; Santhosh, C.; Eswara Varaprasada Rao, K.; Kim, D.K.; Nirmala Grace, A. Enhanced properties of porous $CoFe_2O_4$–reduced graphene oxide

composites with alginate binders for Li-ion battery applications. New J. Chem., 2014, 38, 3654–3661.

[89] Fei, P.; Zhong, M.; Lei, Z.; Su, B. One-pot solvothermal synthesized enhanced magnetic zinc ferrite–reduced graphene oxide composite material as adsorbent for methylene blue removal. Mater. Lett., 2013, 108, 72–74.

[90] Zhang, W.; Quan, B.; Lee, C.; et al. One-step facile solvothermal synthesis of copper ferrite–graphene composite as a high-performance supercapacitor material. ACS Appl. Mater. Interf., 2015, 7, 2404–2414.

[91] Shahnavaz, Z.; Woi, P.W.; Aliasn, Y. A hydrothermally prepared reduced graphene oxide-supported copper ferrite hybrid for glucose sensing. Ceram. Int., 2015, 41, 12710–12716.

[92] Peng, E.; Shi Guang Choo, E.; Chandrasekharan, P.; et al. Synthesis of manganese ferrite/graphene oxide nanocomposites for biomedical applications. Small, 2012, 8(23), 3620–3630.

[93] Yang, Y.; Shi, H.; Wang, Y.; et al. Graphene oxide/manganese ferrite nanohybrids for magnetic resonance imaging, photothermal therapy and drug delivery. J. Biomater. Appl. 2016 Jan;30(6):810–22. http://www.ncbi.nlm.nih.gov/pubmed/26296777

[94] Panwar, R.; Puthucheri, S.; Singh, D.; Agarwala, V. Design of ferrite–graphene-based thin broadband radar wave absorber for stealth application. IEEE Trans. Magnet., 2015, 51(1), 2802804, 4 pp.

[95] Durmus, Z.; Durmus, A.; Kavas, H. Synthesis and characterization of structural and magnetic properties of graphene/hard ferrite nanocomposites as microwave-absorbing material. J. Mater. Sci, 2015, 50, 1201–1213.

[96] Gao, T.; Chen, Z.; Huang, Q.; Niu, F.; Huang, X.; Qin, L.; Huang, Y. A review: preparation of bismuth ferrite nanoparticles and its applications in visible-light induced photocatalyses. Rev. Adv. Mater. Sci., 2015, 40, 97–109.

[97] Dai, J.F.; Xian, T.; Di, L.J.; Yang, H. Preparation of $BiFeO_3$-graphene nanocomposites and their enhanced photocatalytic activities. J. Nanomater., 2013, 642897, 5 pp.

[98] Li, T.; Shen, J.; Li, N.; Ye, M. Hydrothermal preparation, characterization and enhanced properties of reduced graphene-$BiFeO_3$ nanocomposite. Mater. Lett., 2013, 91, 42–44.

[99] Candini, A.; Klyatskaya, S.; Ruben, M.; Wernsdorfer, w.; Affronte, M. Graphene spintronic devices with molecular nanomagnets. Nano Lett., 2011, 11, 2634–2639.

[100] Lopes, M.; Candini, A.; Urdampilleta, M.; et al. Surface-enhanced Raman signal for terbium single-molecule magnets grafted on graphene. ACS Nano, 2010, 4(12), 7531–7537.

[101] Govan, J.; Gun'ko, Yu.K. Recent advances in the application of magnetic nanoparticles as a support for homogeneous catalysts. Nanomaterials, 2014, 4, 222–241.

[102] Lim, C.W.; Lee, I.S. Magnetically recyclable nanocatalyst systems for the organic reactions. Nano Today, 2010, 5, 412–434.

[103] Sivashankar, R.; Sathya, A.B.; Vasantharaj, K.; Sivasubramanian, V. Magnetic composite an environmental super adsorbent for dye sequestration—a review. Environ. Nanotechnol. Monit. Manage., 2014, 1–2, 36–49.

[104] Zhu, J.; Wei, S.; Chen, M.; Gu, H.; et al. Magnetic nanocomposites for environmental remediation. Adv. Powder Technol., 2013, 24, 459–467.

[105] Yin, P.T.; Shah, S.; Chhowalla, M.; Lee, K.-B. Design, Synthesis, and characterization of graphene–nanoparticle hybrid materials for bioapplications. Chem. Rev., 2015, 115(7), 2483–2531.

Chapter 2

Next-Generation Graphene-Based Membranes for Gas Separation and Water Purifications

Quan Xu and Wenwen Zhang

Additional information is available at the end of the chapter

http://dx.doi.org/10.5772/64396

Abstract

Advanced membrane systems are regarded as an important portion of controllable separation processes, such as gas separation and water purification. The ideal materials should have good permeability for selected particle sizes, high stiffness to withstand high pressures applied, large surface area and micro- or nanopore structures for excellent selectivity. Recently, graphene with oxygen-containing functional groups and graphene oxide (GO) nanosheets, obtained via chemical oxidation of graphite, achieved tremendous properties that include excellent mechanical strength, large relative surface area, unique honeycomb lattice two-dimensional structure as well as narrow pore distribution, offering platform to be used as advanced, ultrathin membrane for a wide variety of purification process with high efficiency. In this review chapter, the potential application of such advanced materials for gas separation and water purification process is discussed. The fabrication and modification process and innovation of such advanced two-dimensional functional structure for purification and separation process are introduced. This review chapter will offer opportunity to understand details involved in gas and/or water molecular transport through thin, laminar graphene oxide and derived structures, as well as up to now progress in the field.

Keywords: graphene, graphene oxide, membrane, purification, separation

1. Introduction

Separation, or separation technique, is regarded as an important process that converts a mixture of substances into distinct pure mixtures/fractions. In industry, separation is an important indispensable process for nature gas separation, hydrogen production, seawater desalination, etc. [1–3]. Traditional separation technologies, including distillation [4, 5], adsorption [1], absorption [6, 7], etc., have been developed for decades. With the restriction of total energy resources on Earth and the continuous global population growth, the demand for high efficacy technologies to solve urgent environmental and energy problems urgently needs the formulation of advanced separation techniques. In comparison to traditional technologies, membrane separation technology has been developed and became a well-established separation technology over the past decennia, with the benefit of overall low energy usage, flexible arranged process conductions, safety, materials sustainability and being economical and environmental friendly [8, 9]. Membranes work as a selective separation wall which target substances that can pass through and block others [8, 10, 11]. Two important factors are used to evaluate the performance of membranes: flux and selectivity [12]. Flux is defined as the volumetric flow rate, and selectivity is the status of molecules being passed through. The traditional membranes generally made from polymers, zeolite and silicon have been successfully applied in industry [13]. A summary of membrane porous sizes *versus* application classification is exhibited in **Figure 1**.

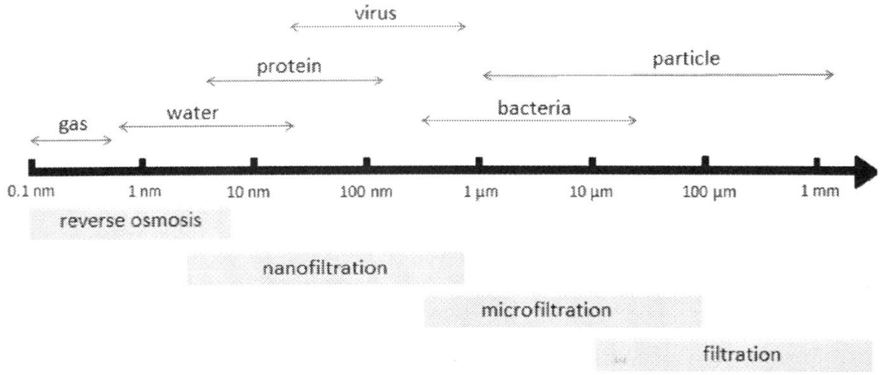

Figure 1. Summary of membrane classification according to their pore sizes.

Recently, carbon-based nanomaterials have attracted intensive attention from scientists, due to their abundant reserves, easy accessibility, remarkable mechanical properties, biocompatibility, as well as environmental friendly properties [14, 15]. Initially, carbon nanotubes (CNTs) were believed to be excellent candidates for membrane materials, due to their uniquely hollow structure with one open-end, and extremely high mechanical performance [16, 17]. However, the fabrication of vertically aligned CNT arrays with high density, large scale quantity and low defects is still challenging [18]. Thus, the generation of CNT-based separation membrane, for large-scale practical applications, remains a theoretical possibility. Another kind of carbon-based nanomaterials with great mechanical properties is carbon diamond. They contain amorphous sp^3-hybridized carbon atoms and are also considered as candidate materials for separation technology. Considered the large energy consumption and high cost, it is not economical and practical to use carbon diamonds for large-scale industrial applications [19].

http://dx.doi.org/10.5772/64396

Therefore, the development of more applicable and economically suited carbon-based nanomaterials for fabricating separation membranes is an urgent need.

The two-dimensional carbon, graphene, possessing a single-atom-thick sheet of sp^2-hybridized carbon atoms arrayed in a honeycomb pattern, opened a new door for researchers and scholars to fabricate the next generation of membranes with outstanding separation capacities [20, 21]. Graphene and its derivative, graphene oxide (GO), with very thin, stiff and large surface area and micro-/nanopore structure layers, are considered as the ideal candidates for membrane materials, due to their high durability, withstanding high applied working pressures and excellent selectivity [22–25]. Graphene is stripped out of graphite materials made of only carbon atoms, with a kind of two-dimensional crystal structures having just one-atom-thick sheet. The concept, graphene-based materials, as well as the preparation, has been explored for over a century. In the early middle of the nineteenth century, Brodie et al. [26] prepared GO and distinguished graphite as "lamellar" and "amorphous". In 1918, Haenni et al. [27] described in detail the properties of GO paper. Then Ruess et al. published the first transmission electron microscopy (TEM) images of a few layers of graphene (between 3 and 10 layers) imaged in 1948. Hummers [28] provided a redox method to prepare GO with outstanding properties. Tsotsis et al. [29] reported that graphene membranes can be used for gas mixture separation. In 2004, Andre Geim, a physicist from the University of Manchester, and his copartner Konstantin Novoselov, successfully isolated single-layer graphene from a sheet of graphite, providing evidence that two-dimensional materials can exist alone in nature, and they shared the 2010 Nobel Prize of Physics for this work [30]. Nowadays, graphene and derived materials have attracted worldwide attention, and the global market for graphene is reported to have reached $9 million in 2014 [22, 31, 32].

The first experimental use of graphene as membrane materials was performed by Bunch et al. [33], who demonstrated that graphene is impermeable to all standard gas atoms including helium. With high crystal quality of carbon atoms, the electron density of graphene aromatic rings can repel most atoms and molecules [34]. Scientists further suggest that nanoporous graphene (NPG), with monatomic layers, high mechanical strength and strong chemical stability, may have a great potential for nature-/biogas separation and seawater distillation [23, 24, 35, 36]. Moreover, its derived material, graphene oxide (GO), was identified as potential filtration/membrane, due to the controllable pore distribution, applicability under high-pressure conditions, simple preparation procedure and the possibility for easy large-scale production. Many attempts were conducted for gas mixture separation and water or liquid molecules purification [36, 37].

In this chapter, the recent applications and research progresses of graphene and GO on separation membrane for gas separation and water purification are discussed. The chapter is divided into three sections. In the first section, the applications of graphene and GO on gas separation will be introduced. The second deals with water filtration. Finally, a summary and outlook of graphene-based separation membranes will be stated, including the recent challenges and the expected further developments in the future.

2. Graphene and GO for gaseous selectivity separation

Comparing with traditional membrane materials, graphene and GO membranes show remarkable properties, such as unique stiffness, hydrophobic characteristics and are easy scale up in various applications [38–41]. The monolayer graphene is known to resist to very small gaseous permeation, including He (helium) [33], as the electron-clouds on the graphene hexagonal rings could hamper the atoms and molecules to allow the transit of small molecules. Graphene is long being considered as an ideal gas barrier material. Although graphene prevents gas to pass through, nanoporous graphene (NPG), due to its high surface porosity, can efficiently allow gas flow to selectively pass through, thus becoming a potential membrane material candidate. The analysis of NPG as membrane materials starts with the simulation of this amazing material. In 2008, Sint et al. [42] designed functionalized nanopores in graphene monolayers and showed that they could provide highly selective passage of hydrated ions, using molecular dynamics (MD) simulations. In the same year, Jiang et al. [43] studied NPG and investigated the permeability and selectivity of NPG and demonstrated the remarkable potential of graphene as a membrane for the gas separation of H_2 and CH_4. Their results indicated that H_2 permeance could be up to 1 mol/$m^2s^1Pa^1$, with a selectivity of amazing 10^8 for H_2/CH_4, which was up to 10^7 improvement in the value obtained with silica membranes (**Figure 2**). Du et al. [44] designed a series of NPG for separating N_2 and H_2 and found that there were different mechanisms for H_2 and N_2 to

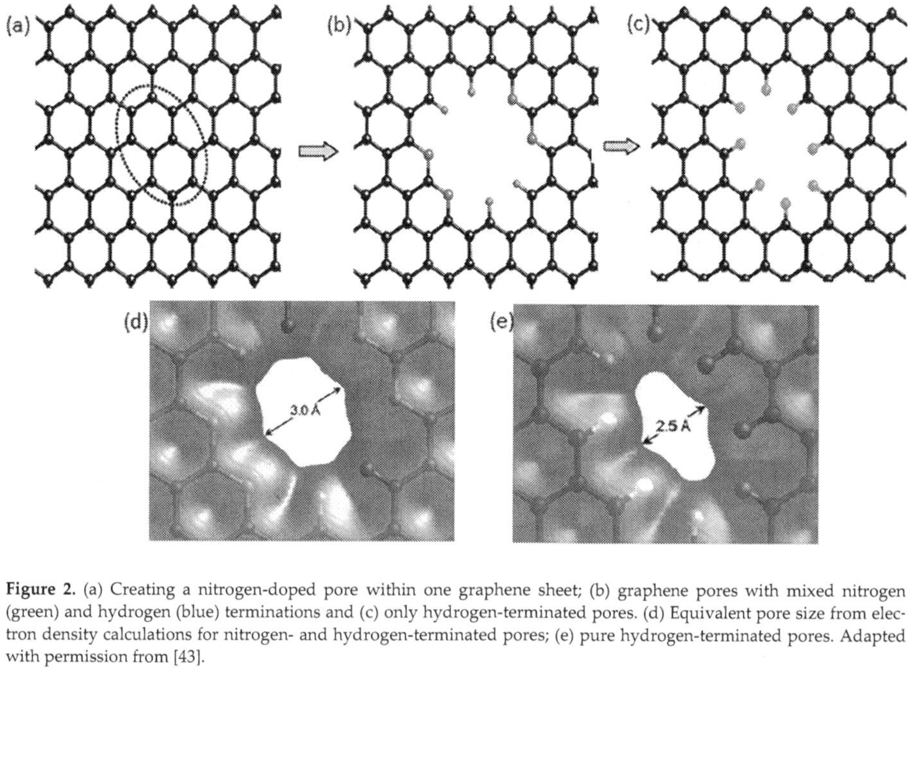

Figure 2. (a) Creating a nitrogen-doped pore within one graphene sheet; (b) graphene pores with mixed nitrogen (green) and hydrogen (blue) terminations and (c) only hydrogen-terminated pores. (d) Equivalent pore size from electron density calculations for nitrogen- and hydrogen-terminated pores; (e) pure hydrogen-terminated pores. Adapted with permission from [43].

http://dx.doi.org/10.5772/64396

permeate through the NPG membrane. The hydrogen flux was linear in relation to the pore size of NPG, while nitrogen flux was not. This revealed that the mechanisms of hydrogen and nitrogen permeation through the porous graphene are different. Trinh et al. [45] calculated bonding energy between gas mixtures including molecules of CO_2 and H_2 and graphite by performing MD simulation, and then analyzed selectivity and self-diffusion of CO_2 and H_2 on graphite membrane layers at different temperatures. **Figure 3** shows a series of typical snapshots, at different temperatures, of mixtures of CO_2 and H_2 on graphite surfaces, and the number of particles, as shown by Trinh et al. [45]. These authors also reported that the binding energy of CO_2 on graphite surface was three times larger than that of H_2, and the separation of CO_2 over H_2 was five times larger at lower temperature, compared to the value obtained at higher temperature. Other mechanisms, such as the pore translocation-limited, surface adsorption and rearrangement reaction, have been in-depth explored and discussed in the literature [46–49].

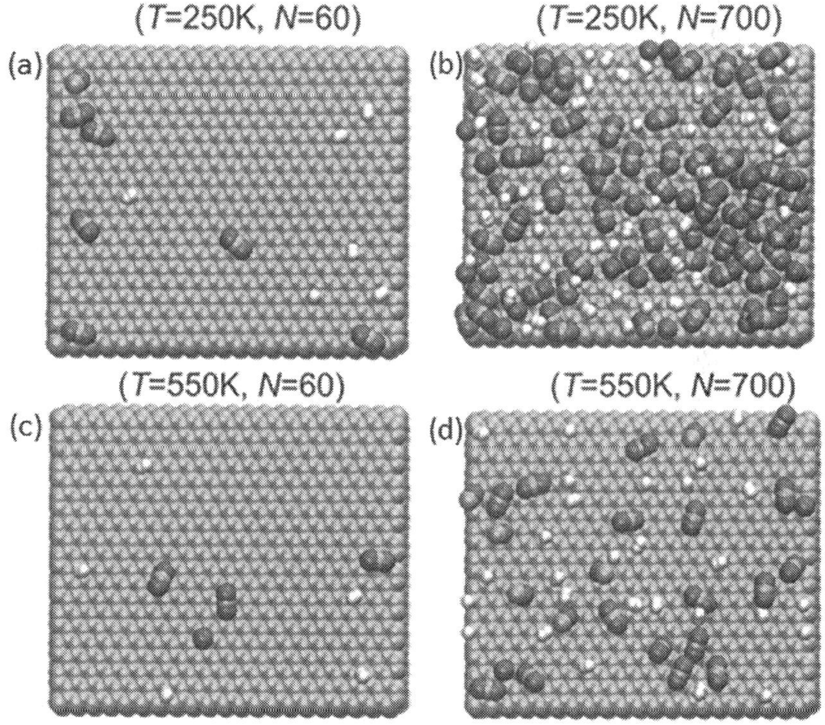

Figure 3. Typical snapshots of molecules absorbed on graphite surface at different temperatures (T) and total numbers of particles (N). The green, red and white represent C, O (oxygen) and H atoms, respectively [45].

Taking theoretical results into consideration, graphene and GO are promising materials to become the next-generation separation membranes. Scientists start to develop different strategies for synthesizing NPG. Most common methods include electron beam [50], ultraviolet-induced oxidative etching [51], electrochemical exfoliation [52] and creation of porous two-dimensional sheets by chemical synthesis [53] or 3D structures by sandwiching/tethering graphene with carbon nanotubes [54]. Fischbein and Drndić [50] introduced the use of e-beam irradiation to create nanopores on graphene sheets, allowing the porous size on the graphene to be controlled. Russo et al. prepared graphene nanopores with just a 3 Å radii, equivalent to 10 atoms removed, which is shown in **Figure 4** [55]. Ultraviolet light possesses energy to break certain chemical bonds and etching holes on graphene sheets. With this method, Bunch et al. [51] successfully prepared micrometer-sized graphene membranes with pores, which allowed smaller molecules to pass through but not larger molecules. Ambrosi and Pumera [52] tested different electrolytes in the method of electrochemically exfoliation, and suggested that graphene was a great candidate for electrochemical applications, such as biosensing and energy storage. O'Hern et al. [56] fabricated graphene composite membranes by transferring large area GO sheets onto polymer substrates. Strategies on how to precisely control the thickness and the chemical functionalization groups of graphene sheets [57] were also introduced. Recently, some novel methods are emerged to prepare graphene membranes. For example, Raghavan et al. [58] used plasma to prepare graphene membranes on the substrates including 6H-SiC etched by CF_4- and Cl_4-based plasmas and the results showed that the etching process could selectively etch silicon to produce carbon-rich surface layers on the silicon carbide substrate. When annealed, these carbon-rich layers formed graphene films with halogen- and oxygen-based defects. Reserbat-Plantey et al. [59] introduced an on-chip nanoelectromechanical system (NEMS) for preparing defect-controllable graphene which was believed to be the process allowing preparation of graphene at room temperature. Zhong et al. [60] reported that graphene sheets modified with the appropriate pore size of 4.06 Å could efficiently separate N_2 and CO_2 by molecular dynamic (MD) simulation and found that the graphene sheets behaved like a mesh that blocked CO_2 permeance but allowed N_2 to pass through. Besides that, Schrier [61] synthesized a porous graphene structure to separate He, Ne and CH_4, and observed that the porous graphene could separate He from alkanes and noble gases. Hauser and Schwerdtfeger [62] discovered that nanopores functionalized graphene membrane with nitrogen-doped porous had the ability to efficiently separate ^{3}He from ^{4}He. Zhang et al. [63] designed covalently cross-linked free-standing few-layered GO films using a layer-by-layer deposition technique, in combination with photochemical cross-links, and achieved a controllable and adjustable graphene membrane with good stability that lasted up to 9 days of use. In 2016, Lozada-Hidalgo et al. [64] used graphene membranes to filtrate hydrogen ion isotopes, like molecular sieves, and found that deuteron permeation through these membranes is much slower than proton permeation. This work implies that graphene has the potential to be used as nuclear waste treatment. With the developed technique, larger high quality and uniform graphene sheets can be transferred to various substrates. Graphene sheets, the ultimate membrane platform, will have wider usages in gas separation technology.

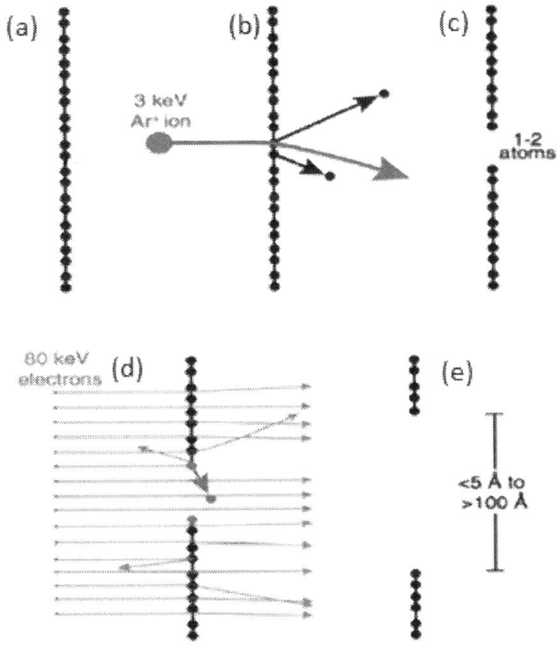

Figure 4. Preparation of graphene nanopores by e-beam. Starting with pristine, one-atom-thick graphene (a), a monatomic site is created by bombardment with 3 keV Ar (Argon)$^+$ ions, which remove a few carbon atoms in a single bombardment (b). After the single-bombardment, a small but stable defect-location is left in the layer surface (c). Then using a uniform e-beam enlargement of the defect-location is achieved by removing the edge atoms (d). Finally, the pore is opened until reaching the preferred size (e) [55].

Compared to graphene, GO membrane materials have a higher fracture energy (350 kJ m^{-3}) and tensile strength (42 ± 2 GPa) [65]. GO can be bent over 90° and form a simple curve (**Figure 5a**) [65]. This stiff-yet-flexible property makes GO an ideal candidate as membrane material. Various routes have been reported for preparing graphene and its oxide derivative with good mechanical properties for different applications. Pierleoni et al. [66] used soluble graphene derivatives to fabricate gas separation membrane coatings on the surface of polymers and found that the use of 2D materials as surface coatings, instead of bulk additives, overcame common issues related to dispersion of graphene into a polymer matrix, and gave a remarkable advantage in preserving the mechanical properties of the bulk polymer. Kim et al. [67] successfully loaded a large area of graphene and GO on the polymer substrate and then measured the transport and selectivity of O_2-N_2 and CO_2-N_2. They considered that the defects and the wrinkles on the graphene and GO surfaces (**Figure 5b**) conferred the selectivity and the controllability to the membranes. Bai et al. [68] invented a new graphene nanostructure named graphene nanomesh (**Figure 5c**). They used block copolymer lithography to prepare NPG with neck widths as low as 5 nm and such materials exhibited great membrane performance. Yang et al. [69] noticed that the wrinkles on the graphene could lead to buckling behavior, which would significantly influence graphene membrane performance. Li et al. [70] reported ultrathin GO membranes with thickness approaching 1.8 nm, loaded on porous anodic aluminum oxide (AAO). The selectivity of the membranes could reach as high as 3400 and 900 for H_2/CO_2 and H_2/N_2, respectively. With successful fabrications of gas flow channels and pore sizes, one can optimize the performances of graphene and GO membranes. Other parameters, including temperature and pressure of the system, can affect the effusing rate, selectivity and flux of graphene and GO membranes. Functional groups inserted at the edge of the pores or between the interlayers can also significantly enhance the selectivity and flux of the graphene-based membranes.

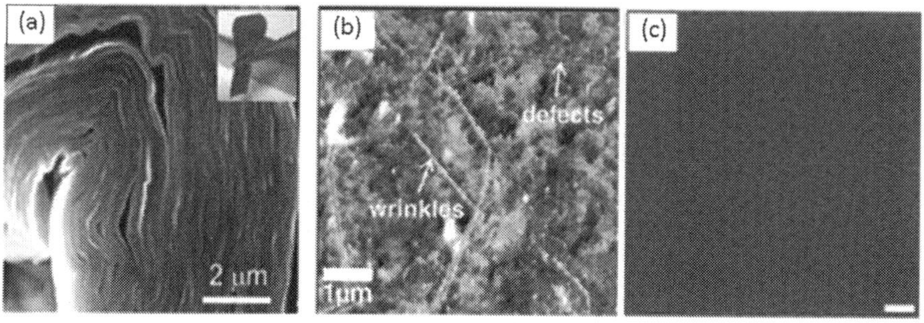

Figure 5. (a) GO materials can be folded over 90° [65]. (b) Atomic force microscopy (AFM) image of graphene membrane surface showing the defects and the wrinkles [67]. (c) Scanning electron microscope (SEM) image of graphene nanomesh [68].

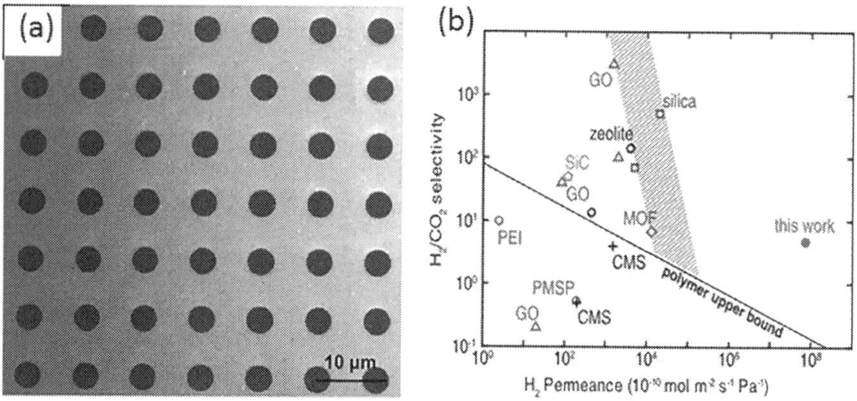

Figure 6. (a) The SEM image of the SiN_x membrane with 49 pores showed. (b) Comparison of hydrogen-carbon dioxide gas mixture separation performances between the graphene membranes and other membranes: GO, poly (1-trimethyl-silyl-1-propyne) (PMSP), polyetherimide (PEI), carbon molecular sieve (CMS), zeolite, silica, metal-organic framework (MOF) and SiC [71].

Traditional membranes, including silica-, polymer- and zeolite-based, follow the well-known law that selectivity is determined by the product of diffusivity and solubility [13]. However, this rule is no longer valid for graphene- and GO-based membranes. For atom-thick graphene, the selectivity is determined by the pore size of GO. The thickness of graphene or GO is also no longer linearly related to the diffusion rates. Although the mechanism is still under intensive study, NPG has shown fluxes of gas in far excess, compared to that of traditional membranes. For example, Celebi et al. [71] fabricated highly efficient mass transfer porous bilayer NPG onto a SiN_x frame punctured with matrix pores with 4 μm in diameter (**Figure 6a**), which physically perforated double-layer graphene. The results (**Figure 6b**) showed that the permeation of the NPG membranes is much higher than that of other traditional membranes, including polymer, silicon and composite materials.

3. Graphene and GO for water purification

Oceans contain about 97% of the world's water. Nowadays the cost of reverse osmosis (RO) decreased to 1.8 KWh/m^3, which is only 1/3 of the cost as was in 1990 [9]. Thermal desalination, such as multiple-effect distillation and multistage flash, needs much more energy to be used, compared to RO. Thus, the development of desalination can only become a sustainable option for water supply if a dramatically new technology is developed.

Graphene, with its extraordinary properties and environmentally friendly nature, has attracted the scientists' interest due to its potential applications in wastewater treatment and water desalination. As a single layer of graphene is impermeable to gases and liquids, NPG is considered as a potential membrane material for molecular sieving or water filtration. Advantages of NPG over RO membranes include negligible thickness and high mechanical strength, indicating faster water flux, with low pressure requirements. The first simulation work on water flux across graphene membrane was conducted by Suk et al. [72]. As shown in **Figure 7**, these authors compared the water flux performances of NPG and CNTs. Results indicated that NPG had higher water flux which did not significantly vary with the thickness of NPG. Then, Cohen-Tanugi and Grossman [73] showed that nanometer-scale pores in single-layer freestanding graphene could effectively filter NaCl salt from water through MD simulation results. They demonstrated that NPG membrane had the ability to prevent salt passage, critically depending on the pore diameter and chemical functional groups bonded to the edges of graphene pores. Nair et al. [74] indicated anomalous water permeation to low friction between the water monolayer and pristine graphene regions. Based on that assumption, a capillary-driven flow mechanism was established for graphene-based membranes. Boukhvalov et al. [75] analyzed the hybrid systems consisting of water and GO and believed formation of hexagonal ice bilayer in between the flakes, as well as melting transition of ice at the edges of flakes, were crucial to understand the water permeation across GO. Ma et al. [76] discovered the fast diffusion of water nanodroplets on graphene and opened a prospect for achieving fast and controllable motion of adsorbates across graphene and GO surface materials.

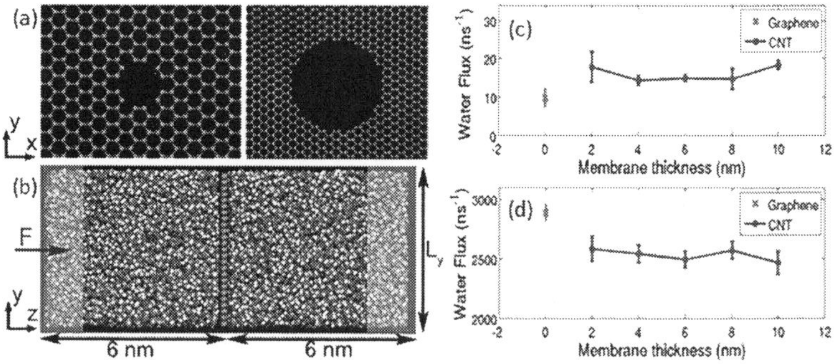

Figure 7. (a) Graphene membrane with nanopore diameter: D_{avg} = 0.75 nm (left) and D_{avg} = 2.75 nm (right). (b) Simulation setup. Cyan color represents carbon atoms; red and white colors represent oxygen and hydrogen atoms of a water molecule, respectively. Two water reservoirs are attached to each side of the porous graphene membrane. Ly=Lx=4 nm when the pore diameter is 0.75 nm, and Ly = Lx = 6 nm when the pore diameter is 2.75 nm. In the shaded region (Δz =1 nm), external forces are applied on water molecules to create a pressure drop across the membrane. (c) Water flux through a (10, 0) CNT with a diameter of 0.78 nm and through a graphene nanopore with a diameter of 0.75 nm. In both cases, a single-file water structure is observed. (d) Water flux through a (20, 20) CNT with a diameter of 2.71 nm and through a graphene nanopore with a diameter of 2.75 nm. A single-file water structure is not observed in this case [72].

In order to improve water permeability, small sheet size, high density of interedge spaces and wide nanochannels are required in graphene and GO membrane fabrication. Wei et al. [77] enhanced water flow in graphene membranes into the unique porous microstructures. These authors pointed out that the side-pinning effects caused by H-bonds between water molecules and oxidized regions could help fast water transportation in graphene channels, and that this, along the porous microstructure, were the principal factors responsible for the enhanced water flow. Other studies also indicated that intercalation of different-sized cross-linkers can tune the interlayer channel size [78]. Furthermore, Wei et al. [79] investigated interlayer flow and demonstrated that the breakdown of flow enhancement could be derived from GO chemical functionalization as well as relaxation of nanoconfinement. Also, both simulation and experiment were performed to evaluate the characteristics of membrane structure and operation parameter affect water flux in graphene and GO-based membranes. Cohen-Tanugi and Grossman [73] found out that the water flux through monolayer graphene had a linear relationship with the pore area by computational methods. They also proposed that hydrophilic pores could increase water permeation, due to H-bonding between water molecules and pore edges. Hu and Mi [80] fabricated a series of novel GO membranes with different numbers of layers and tested their water purification performance. These authors observed that there was no evident correlation between water flux and the number of layers of graphene. Sheath and Majumder [81] prepared graphene membranes by evaporation-induced capillary-force self-assembly and found out that membranes had higher space rejection properties, while simultaneously enhancing water flux, compared to analogous materials produced by the traditional vacuum filtration technique. Han et al. [82] synthesized ultrathin graphene nanofiltration membranes (uGNMs) on nanoporous microstructure substrates, using chemically converted graphene. The uGNMs showed high retention (>99%) for organic dyes and moderate retention (≈20–60%) for ion salts. The physical sieving and electrostatic interaction were considered as two important factors dominating the rejection process (**Figure 8**).

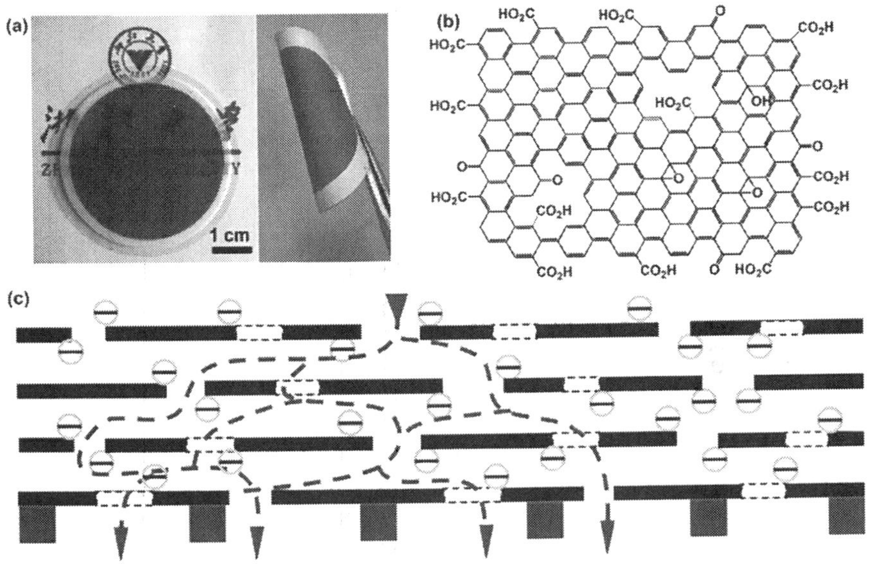

Figure 8. (a) Digital photo of an uGNM coated on an AAO disk (left) and a twisted uGNM coated on a PVDF membrane (right). (b) Schematic representation of a brGO: graphene sheet with a certain amount of holes; most of the oxidized groups are located on the edges and the periphery of the holes. Note that the real graphene sheets extend further than depicted. (c) Schematic view of possible permeation route: water molecules go through the nanochannels of the uGNMs and holes of the graphene sheets and finally reach the pores of supporting membranes. The blank squares present the holes on the graphene sheets (black line). The edges of the brGO and the periphery of the holes are negatively charged [82].

In order to further optimize the performances of NPG or GO for water treatment, the mechanism involved needs to be intensively studied. There are two main primary factors concerning the selectivity of the materials used in this process: size exclusion and interactions with functional groups, including chemical and electrostatic ones. Using classical molecular dynamics, Cohen-Tanugi and Grossman [73] concluded that nanopores in single-layer freestanding graphene could effectively filter NaCl salt from water. Small pores, low pressure and hydrophobic pores refuse salts more efficiently, due to direct size exclusion, while the larger effective volume of ions and the lack of H-bonding can lead to a higher energy barrier to ions passage. Prince et al. [83] devised an ultrawetting graphene-based membrane by incorporating amine and carboxyl functionalities onto graphene and then covalently attaching it to a polymer matrix to fabricate a water filtration membrane. They concluded that the graphene increased the water permeability of the membrane by 126% compared with stable membrane selectivity. Hu and Mi [84] prepared a kind of water separation membrane using GO nanosheets and this membrane displayed remarkable rejection of monovalent and divalent salts and moderate or high rejection of organic dyes. Yin et al. [85] prepared a thin-film nanocomposite (TFN) membrane containing GO sheets to enhance polyamide (PA) membranes for water purification application. These GO sheets were well dispersed in the polyamide (PA) layer and their incorporation improved membrane performances. Authors suggested that interlayer spacing of GO may serve as water channels and hence contribute to the water permeability enhancement. Independently, Han et al. [82] obtained similar results, showing that the observed rejection organic dyes could be assigned to size, electrostatic interaction, and the rejection of ions is based on Donnan's exclusion (or Donnan effect, a name for the behavior of charged particles near a semipermeable membrane that sometimes fail to distribute evenly across the two sides of the membrane). Additionally, Huang et al. [20] investigated different factors including salt concentration, pH and pressure effects on the separation performance of GO membrane toward small molecules and demonstrated how these factors affect such process. Sun et al. [86] confirmed the selection ion penetration and water purification properties of GO membranes. They reported that sodium salts could quickly pass through GO membranes, while heavy-metal salts moved slower (**Figure 9**). Copper salts were absolutely unable to permeate through, as well as organic contaminants. Their follow-up work [87] revealed, by theoretical calculations, that the coordinative interactions play an important role on the selectivity of the transmembrane transport of alkali and alkaline earth cations. Joshi et al. [88] designed an experiment (**Figure 10a**) to measure the permeation properties of micrometer-thick GO laminates. The normalized permeation rates as per 1 M feed solution (**Figure 10b**) show that ultrafast permeation is only possible for ions smaller than a specific size. Size exclusion affected the selectivity as well as capillary forces could induce the anomalous fast transportation. Their follow-up work [89] indicated that graphene had a large possibility to be used as membranes for desalination and confirmed that 100% salt rejection could be achieved for commonly used ions by utilizing single-layer nonporous graphene. Liu et al. [90] studied the pressure-driven water transport inside the nanochannels of graphene membranes. They found that an ultrafast water flux was observed inside the channels, which increased linearly with the driving pressure, but decreased as the interaction strength between the water molecules and the bilayer increased.

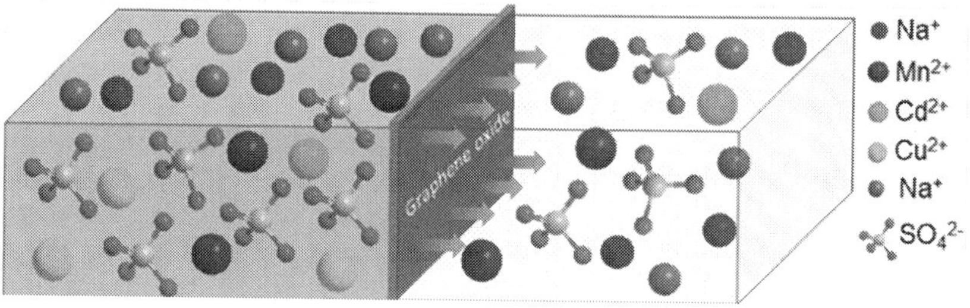

Figure 9. Schematic diagram of the penetration processes of different ions through GO membranes [86].

http://dx.doi.org/10.5772/64396

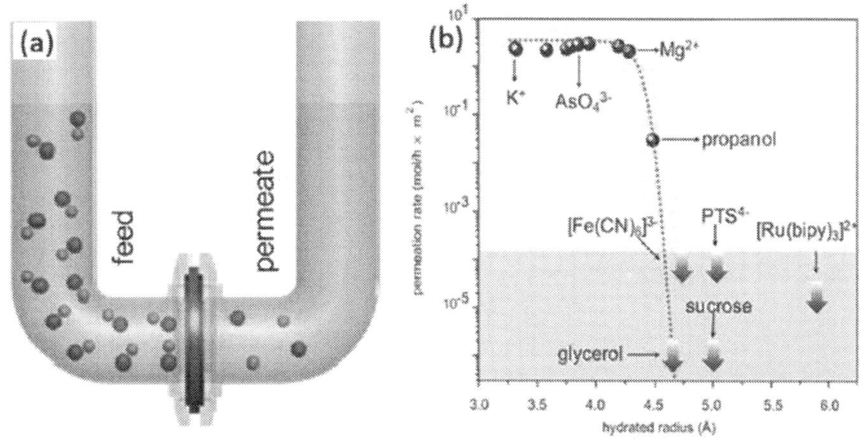

Figure 10. (a) Experimental setup. A U-shaped tube is partitioned by GO membrane into two compartments that are referred to as feed and permeate. (b) Sieving through atomic-scale mesh. The small species permeate with approximately the same speed, whereas large metal ions and organic molecules exhibit no detectable permeation [88].

Apart from water dissipation and purification, graphene and GO can also be used on other water treatment applications, such as recovery of acids from iron-based electrolytes. Sun et al. [91] reported that nano-GO (**Figure 11**) membranes could completely block Fe^{3+} when the ion concentration reached a certain extent (0.01 mol L^{-1}), whereas the H^+ permeated through unimpeded. This implies that the use of GO membranes can be applied to the recovery of acids from an $FeCl_3$ solution. On the other hand, Huang et al. [21] prepared a GO membrane supported on a ceramic hollow fiber for separating dimethyl carbonate–water mixtures through a pervaporation process. They considered that the separation process followed a sorption-diffusion mechanism. Tang et al. [92] also utilized a pressurized ultrafiltration for dehydration of ethanol using GO membranes. However, the results obtained from binary-component feed tests were lower than the ideal water/ethanol selectivity, found by calculation. This may be attributed to the effects of intermolecular H-bonding between water molecules and the functional groups of GO nanosheets that may enlarge the interlayer spacing and allow more ethanol transport through the GO membrane. Recently, Lin et al. [93] devised a way to fast produce graphene in a CVD process and then grow it on Ni, Cu and a Cu-Ni alloy with proper composition. This work provided a possibility to produce large-scale graphene single crystals, which paved the route towards reaching practical applications of graphene, such as in wafer-scale electronic and optoelectronic devices. Wu et al. [94] used a similar approach to fast grow inch-sized single-crystalline graphene by letting a single nucleus interact with a monolayer at a fast rate. They synthesized a roughly 1.5-inch-large graphene one-atom-layer in 2.5 h and suggested that this method may prove effective for the synthesis of wafer-scale single-crystalline monolayers of other two-dimensional materials.

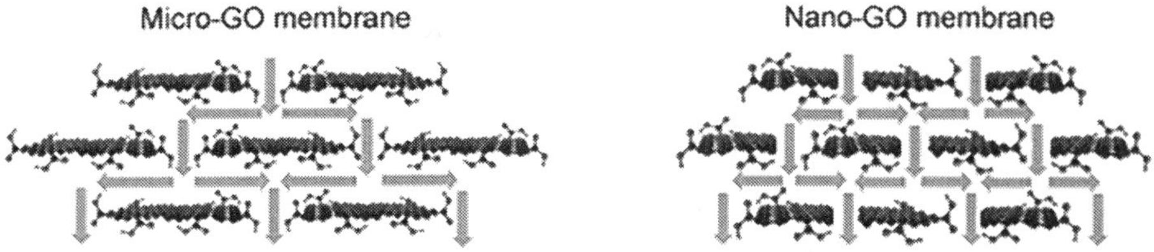

Figure 11. Schematic diagrams of the structure of nanocapillary networks in nano- and micro-GO membranes. It can speculated that when the lateral dimensions of the GO sheets used to form the membranes were reduced from micrometer (left) to nanometer (right), the amount of nanocapillaries formed within the membranes should significantly increase, resulting in the enhancement of ion penetrations through GO membranes [91].

In summary, graphene and GO membranes are believed to be the next-generation separation materials for applications in water purification due to their significant intrinsic mechanical strength [95], high chemical stability [57], high antibacterial activity [96] and exquisite antifouling properties [65], as mentioned above. The narrow pore size and nanochannels of GO membranes can be tuned by induction and modification of surface functional groups. For water separation and purification, an ideal pore size distribution should be achieved as this is beneficial to precisely sieving ions through the size exclusion mechanism.

4. Conclusions and perspective

Graphene and its oxide derivative, having two-dimensional one-atomic-layer, ultrafast permeation, remarkable mechanical properties and outstanding energy-efficiency, are promising candidates for the next-generation separation membranes for precise and selective gas separation (**Figure 12**). Although graphene and GO have been reported as being extraordinary in both selectivity and flux when separating multiple gas mixtures and water purifications, there are still several challenges that need to be solved in order to achieve industrial applications. Firstly, the pores on the surface of graphene and GO sheets can be blocked by defects, impurities, polymer, molecular residues and grain boundaries. They will affect the performance of the membranes, namely the flux rate [97]. Therefore, the mixture should be carefully purified before entering into the graphene membrane systems. The operating conditions, such as pressure, temperature and flux rate, will affect membrane selectivity. These factors must be considered before using graphene- and GO-based membrane materials. Secondly, the selectivity for H_2/H_2S, CO_2/CH_4 or mixtures of three or more gases on gas separation applications of graphene and GO membranes still needs to be fully characterized. A reasonably analysis of the parameters that influence the process of graphene and GO is needed for predicting the mechanism of membrane selectivity. Those information are of great interest for industries, such as nature and shale gas companies. Although graphene is considered as the world's strongest material with Young's modulus up to 1 TPa, local defects, cracks and wrinkling of graphene and GO will affect the excellent property of graphene sheets [98]. The local bond breaking and bond rotation at the crack tip of the graphene could lead to dramatic losses of mechanical properties of corresponding products, which may cause the membranes to be less robust and more fragile, and even generate unexpected large pores during constant usage. To solve this issue, a good solution might be to use a reliable porous substrate as support structure for graphene membranes. The connection and adhesion properties between membranes and support structures thus need to be fully revealed. Then, for the membrane itself, pore size is the most critical parameter in controlling the performance of graphene-based membranes; graphene and GO can enhance the selectivity of separating mixtures by adjusting its layer-to-layer distance and/or inserting functional groups on the surfaces and interlayer-faces of graphene sheet. The last challenge is the overall cost for the industry demands. Large-scale graphene and GO membrane fabrication techniques need to be developed, and related standard quality control tests need to be created to meet the industry requirements. Although the high production cost of GO so far impeded the development of large-scale applications, the coming advances on technology will certainly decrease the manufacturing prices of graphene and GO, and more methods will be developed to generate defect-free, larger area graphene, with controllable pore structure/size and interlayer distance. It is believed that this promising field will attract increasing interest from both academia and industry in the future.

Water purification is also a promising topic worth exploring, in order to achieve the fundamental understanding that might lead to the specialized industrial application of graphene- and GO-based membranes. The first challenge is the precise design of user-controlled GO membranes, including the pattern of oxidized regions, interlayer space, pore size, modification of pore chemistry and the number of graphene layers. Further developments on the parameters that determine the small molecule permeation and separation performance are also necessary. Meanwhile, the effect of the interlayer space and surface defect of graphene and GO on fluid flux is being carefully studied. With those information, the mechanism affecting the selectivity of NPG and GO may be solved. The contradiction between flux and selectivity is considered to determine the optimal membrane fabrication parameters. Modification of the surface of GO membranes with different atoms and functional groups, to adjust water affinity, can achieve higher flux, without sacrificing selectivity of the membrane system. The second topic that needs to be considered is the state of GO in aqueous conditions. The swelling needs to be regulated in order to achieve the ideal selectivity. For industrial processes, the operating pressure is an important factor to both separation system and membrane materials. Operation in a large range of pressures will be highly beneficial. Finally, the last issue is the cost. For some special industrial applications, such as producing high purity acid from high-concentration $FeCl_3$ solutions, pretreatment or posttreatment will increase energy consumption and total investment. Researchers are also suggested to minimize the gap between theoretical and experimental results on water/ion transport mechanisms across GO nanosheets in water and to solve impurity issue on the graphene- and GO-based membranes, including adsorption of metals and organic dyes [99], ionic clogging inside the capillaries, as well as coordination occupation on the surface of oxidized regions. These effects still need investigation and optimization. The key to successfully fabricate adequate membrane systems is to balance the antimicrobial efficiency and mechanical strength of the freestanding membranes.

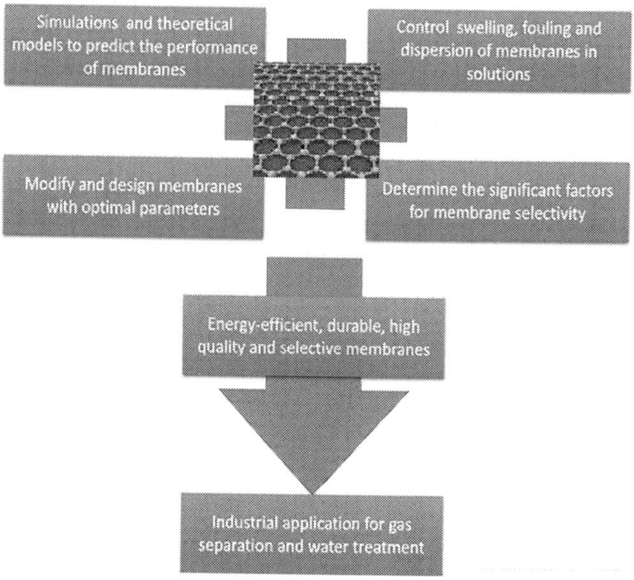

Figure 12. The outlook of the research on graphene and GO membranes.

Author details

Quan Xu[1*] and Wenwen Zhang[2]

*Address all correspondence to: xuquan@cup.edu.cn

1 State Key Laboratory of Heavy Oil Processing, China University of Petroleum, Beijing, China

2 College of Textile, North Carolina State University, North Carolina, USA

References

[1] Geankoplis C. Transport Processes and Separation Process Principles (Includes Unit Operations), Fourth Edition. Upper Saddle River, NJ, USA: Prentice Hall Press; 2003.

[2] Baker RW. Research needs in the membrane separation industry: looking back, looking forward. Journal of Membrane Science 2010;362:134–136.

[3] Baker RW and Lokhandwala K. Natural gas processing with membranes: an overview. Industrial & Engineering Chemistry Research 2008;47:2109–2121.

[4] Earle MJ, Esperança JMSS, Gilea MA, Canongia Lopes JN, Rebelo LPN, Magee JW, et al. The distillation and volatility of ionic liquids. Nature. 2006;439(7078):831–834.

[5] Gao X, Chen J, Ma Z and Yang L. Simulation and optimization of distillation processes for separating a close-boiling mixture of n-butanol and isobutanol. Industrial & Engineering Chemistry Research 2014;53:14440–14445.

[6] Sistla YS, Khanna A. Carbon dioxide absorption studies using amine-functionalized ionic liquids. Journal of Industrial and Engineering Chemistry. 2014;20(4):2497–2509.

[7] Henley EJ, Seader JD, Roper DK. Separation Process Principles: Wiley; 2011.

[8] Scholes CA, Stevens GW, Kentish SE. Membrane gas separation applications in natural gas processing. Fuel 2012;96:15–28.

[9] Elimelech M, Phillip WA. The future of seawater desalination: energy, technology, and the environment. Science. 2011;333(6043):712–717.

[10] Xu Q, Xu H, Chen J, Lv Y, Dong C, Sreeprasad TS. Graphene and graphene oxide: advanced membranes for gas separation and water purification. Inorganical Chemistry Frontier. 2015;2:417–424.

[11] Pabby AK, Rizvi SSH, Requena AMS. Handbook of Membrane Separations: Chemical, Pharmaceutical, Food, and Biotechnological Applications, Second Edition: CRC Press; 2008.

[12] Scott K. Industrial Membrane Separation Technology: Springer Science & Business Media; 1996.

[13] Bernardo P, Drioli E, Golemme G. Membrane gas separation: a review/state of the art. Industrial and Engineering Chemistry Research. 2009;48(10):4638–4663.

[14] Dong C, Campell AS, Eldawud R, Perhinschi G, Rojanasakul Y, Dinu CZ. Effects of acid treatment on structure, properties and biocompatibility of carbon nanotubes. Applied Surface Science 2013;264:261–268.

[15] Campbell AS, Dong C, Dordick JS, Dinu CZ. BioNano engineered hybrids for hypochlorous acid generation. Process Biochemistry. 2013;48(9):1355–1360.

[16] Kemnade N, Shearer CJ, Dieterle DJ, Cherevan AS, Gebhardt P, Wilde G, et al. Non-destructive functionalisation for atomic layer deposition of metal oxides on carbon nanotubes: effect of linking agents and defects. Nanoscale. 2015;7(7):3028–3034.

[17] Baughman RH, Zakhidov AA, de Heer WA. Carbon nanotubes – the route toward applications. Science. 2002;297(5582):787–792.

[18] Yu M, Funke HH, Falconer JL, Noble RD. High density, vertically-aligned carbon nanotube membranes. Nano Letters. 2009;9(1):225–229.

[19] Karan S, Samitsu S, Peng X, Kurashima K, Ichinose I. Ultrafast viscous permeation of organic solvents through diamond-like carbon nanosheets. Science. 2012;335(6067): 444–447.

[20] Huang H, Ying Y, Peng X. Graphene oxide nanosheet: an emerging star material for novel separation membranes. Journal of Materials Chemistry A. 2014;2(34):13772–13782.

[21] Huang K, Liu G, Lou Y, Dong Z, Shen J, Jin W. A graphene oxide membrane with highly selective molecular separation of aqueous organic solution. Angewandte Chemie International Edition. 2014;53(27):6929–6932.

[22] Novoselov KS, Geim AK, Morozov SV, Jiang D, Zhang Y, Dubonos SV, et al. Electric field effect in atomically thin carbon films. Science. 2004;306(5696):666–669.

[23] Tsetseris L, Pantelides ST. Graphene: an impermeable or selectively permeable membrane for atomic species? Carbon 2014;67:58–63.

[24] Gai J-G, Gong X-L, Wang W-W, Zhang X, Kang W-L. An ultrafast water transport forward osmosis membrane: porous graphene. Journal of Materials Chemistry A. 2014;2(11):4023.

[25] Liu H, Dai S, Jiang D-E. Permeance of H_2 through porous graphene from molecular dynamics. Solid State Communications. 2013;175–176:101–105.

[26] Brodie BC. On the atomic weight of graphite. Philosophical Transactions of the Royal Society of London 1859;149:249–259.

[27] Kohlschütter V, Haenni P. Zur Kenntnis des Graphitischen Kohlenstoffs und der Graphitsäure. Zeitschrift für Anorganische und Allgemeine Chemie. 1919;105(1):121–144.

[28] Hummers Jr WS, Offeman RE. Preparation of graphitic oxide. Journal of the American Chemical Society. 1958;80(6):1339–1340.

[29] Sedigh MG, Onstot WJ, Xu L, Peng WL, Tsotsis TT, Sahimi M. Experiments and simulation of transport and separation of gas mixtures in carbon molecular sieve membranes. The Journal of Physical Chemistry. A. 1998;102(44):8580–8589.

[30] Geim AK, Novoselov KS. The rise of graphene. Nature Materials. 2007;6(3):183–191.

[31] Geim AK. Graphene: status and prospects. Science. 2009;324(5934):1530–1534.

[32] Pantelides ST, Puzyrev Y, Tsetseris L, Wang B. Defects and doping and their role in functionalizing graphene. MRS Bulletin. 2012;37(12):1187–1194.

[33] Bunch JS, Verbridge SS, Alden JS, van der Zande AM, Parpia JM, Craighead HG, et al. Impermeable atomic membranes from graphene sheets. Nano Letters. 2008;8(8):2458–2462.

[34] Leenaerts O, Partoens B, Peeters FM. Graphene: a perfect nanoballoon. Applied Physics Letters. 2008;93(19):193107.

[35] Chandra V, Park J, Chun Y, Lee JW, Hwang I-C, Kim KS. Water-dispersible magnetite-reduced graphene oxide composites for arsenic removal. ACS Nano. 2010;4(7):3979–3986.

[36] Liu R, Arabale G, Kim J, Sun K, Lee Y, Ryu C, et al. Graphene oxide membrane for liquid phase organic molecular separation. Carbon 2014;77:933–938.

[37] Goh K, Setiawan L, Wei L, Si R, Fane AG, Wang R, et al. Graphene oxide as effective selective barriers on a hollow fiber membrane for water treatment process. Journal of Membrane Science 2015;474:244–253.

[38] Tao Y, Xue Q, Liu Z, Shan M, Ling C, Wu T, et al. Tunable hydrogen separation in porous graphene membrane: first-principle and molecular dynamic simulation. ACS Applied Materials & Interfaces. 2014;6(11):8048–8058.

[39] Lei G, Liu C, Xie H, Song F. Separation of the hydrogen sulfide and methane mixture by the porous graphene membrane: effect of the charges. Chemical Physics Letters 2014;599:127–132.

[40] Ambrosetti A, Silvestrelli PL. Gas separation in nanoporous graphene from first principle calculations. The Journal of Physical Chemistry C. 2014;118(33):19172–19179.

[41] Mahmoud KA, Mansoor B, Mansour A, Khraisheh M. Functional graphene nanosheets: the next generation membranes for water desalination. Desalination 2015;356:208–225.

[42] Sint K, Wang B, Král P. Selective ion passage through functionalized graphene nanopores. Journal of American Chemical Society. 2008;130(49):16448–16449.

[43] Jiang D-E, Cooper VR, Dai S. Porous graphene as the ultimate membrane for gas separation. Nano Letters. 2009;9(12):4019–4024.

[44] Du H, Li J, Zhang J, Su G, Li X, Zhao Y. Separation of hydrogen and nitrogen gases with porous graphene membrane. The Journal of Physical Chemistry C. 2011;115(47):23261–23266.

[45] Trinh TT, Vlugt TJH, Hägg M-B, Bedeaux D, Kjelstrup S. Selectivity and self-diffusion of CO2 and H2 in a mixture on a graphite surface. Frontiers in Chemistry. 2013;1.

[46] Drahushuk LW, Strano MS. Mechanisms of gas permeation through single layer graphene membranes. Langmuir. 2012;28(48):16671–16678.

[47] Eigler S, Dotzer C, Hirsch A, Enzelberger M, Müller P. Formation and decomposition of CO_2 intercalated graphene oxide. Chemistry of Materials. 2012;24(7):1276–1282.

[48] Sun C, Boutilier MSH, Au H, Poesio P, Bai B, Karnik R, et al. Mechanisms of molecular permeation through nanoporous graphene membranes. Langmuir. 2014;30(2):675–682.

[49] Zhou S, Bongiorno A. Density functional theory modeling of multilayer "epitaxial" graphene oxide. Accounts of Chemical Research. 2014;47(11):3331–3339.

[50] Fischbein MD, Drndić M. Electron beam nanosculpting of suspended graphene sheets. Applied Physics Letters. 2008;93(11):113107–1.

[51] Koenig SP, Wang L, Pellegrino J, Bunch JS. Selective molecular sieving through porous graphene. Nature Nanotechnology. 2012;7(11):728–732.

[52] Ambrosi A, Pumera M. Electrochemically exfoliated graphene and graphene oxide for energy storage and electrochemistry applications. Chemistry – A European Journal. 2016;22(1):153–159.

[53] Kuhn P, Forget A, Su D, Thomas A, Antonietti M. From microporous regular frameworks to mesoporous materials with ultrahigh surface area: dynamic reorganization of porous polymer networks. Journal of American Chemical Society. 2008;130(40): 13333–13337.

[54] Niu J, Li M, Choi W, Dai L, Xia Z. Growth of junctions in 3D carbon nanotube-graphene nanostructures: a quantum mechanical molecular dynamic study. Carbon 2014;67:627–634.

[55] Russo CJ, Golovchenko JA. Atom-by-atom nucleation and growth of graphene nanopores. Proceedings of the National Academy of Sciences. 2012;109(16):5953–5957.

[56] O'Hern SC, Stewart CA, Boutilier MSH, Idrobo J-C, Bhaviripudi S, Das SK, et al. Selective molecular transport through intrinsic defects in a single layer of CVD graphene. ACS Nano. 2012;6(11):10130–10138.

[57] Dreyer DR, Park S, Bielawski CW, Ruoff RS. The chemistry of graphene oxide. Chemical Society Reviews. 2010;39(1):228–240.

[58] Raghavan S, Denig TJ, Nelson TC, Chaudhari S, Stinespring CD. Characterization of graphene-on-insulator films formed using plasma based surface chemistry. Carbon 2016;99:212–221.

[59] Reserbat-Plantey A, Schädler KG, Gaudreau L, Navickaite G, Güttinger J, Chang D, et al. Electromechanical control of nitrogen-vacancy defect emission using graphene NEMS. Nature Communications 2016;7:10218.

[60] Wang Y, Yang Q, Li J, Yang J, Zhong C. Exploration of nanoporous graphene membranes for the separation of N_2 from CO_2: a multi-scale computational study. Physical Chemistry Chemical Physics. 2015.

[61] Schrier J. Helium separation using porous graphene membranes. The Journal of Physical Chemistry Letters. 2010;1(15):2284–2287.

[62] Hauser AW, Schwerdtfeger P. Nanoporous graphene membranes for efficient 3He/4He separation. The Journal of Physical Chemistry Letters. 2012;3(2):209–213.

[63] Huang T, An Q, Luan X, Zhang Q, Zhang Y. Free-standing few-layered graphene oxide films: selective, steady and lasting permeation of organic molecules with adjustable speeds. Nanoscale. 2016;8:2003–2010.

[64] Lozada-Hidalgo M, Hu S, Marshall O, Mishchenko A, Grigorenko AN, Dryfe RAW, et al. Sieving hydrogen isotopes through two-dimensional crystals. Science. 2016;351(6268):68–70.

[65] Dikin DA, Stankovich S, Zimney EJ, Piner RD, Dommett GHB, Evmenenko G, et al. Preparation and characterization of graphene oxide paper. Nature. 2007;448(7152):457–460.

[66] Pierleoni D, Xia ZY, Christian M, Ligi S, Minelli M, Morandi V, et al. Graphene-based coatings on polymer films for gas barrier applications. Carbon 2016;96:503–512.

[67] Kim HW, Yoon HW, Yoon SM, Yoo BM, Ahn BK, Cho YH, et al. Selective gas transport through few-layered graphene and graphene oxide membranes. Science. 2013;342(6154):91–95.

[68] Bai J, Zhong X, Jiang S, Huang Y, Duan X. Graphene nanomesh. Nature Nanotechnology. 2010;5(3):190–194.

[69] Yang K, Chen Y, Pan F, Wang S, Ma Y, Liu Q. Buckling behavior of substrate supported graphene sheets. Materials. 2016;9(1):32.

[70] Li H, Song Z, Zhang X, Huang Y, Li S, Mao Y, et al. Ultrathin, molecular-sieving graphene oxide membranes for selective hydrogen separation. Science. 2013;342(6154): 95–98.

[71] Celebi K, Buchheim J, Wyss RM, Droudian A, Gasser P, Shorubalko I, et al. Ultimate permeation across atomically thin porous graphene. Science. 2014;344(6181):289–292.

[72] Suk ME, Aluru NR. Water transport through ultrathin graphene. The Journal of Physical Chemistry Letters. 2010;1(10):1590–1594.

[73] AU: Please note that Refs. 73 and 80 were identical. Hence, Ref. 80 has been removed and the other references were renumbered in the body of the text and in the reference list.Cohen-Tanugi D, Grossman JC. Water desalination across nanoporous graphene. Nano Letters, 2012;12(7):3602–3608.

[74] Nair RR, Wu HA, Jayaram PN, Grigorieva IV, Geim AK. Unimpeded permeation of water through helium-leak-tight graphene-based membranes. Science. 2012;335(6067): 442–444.

[75] Boukhvalov DW, Katsnelson MI, Son Y-W. Origin of anomalous water permeation through graphene oxide membrane. Nano Letters 2013;13(8):3930–3935.

[76] Ma M, Tocci G, Michaelides A, Aeppli G. Fast diffusion of water nanodroplets on graphene. Nature Materials. 2015;15(1):66–71.

[77] Wei N, Peng X, Xu Z. Understanding water permeation in graphene oxide membranes. ACS Applied Materials & Interfaces. 2014;6(8):5877–5883.

[78] Mi B. Graphene oxide membranes for ionic and molecular sieving. Science. 2014;343(6172):740–742.

[79] Wei N, Peng X, Xu Z. Breakdown of fast water transport in graphene oxides. Physical Review E. 2014;89(1).

[80] Hu M, Mi B. Layer-by-layer assembly of graphene oxide membranes via electrostatic interaction. Journal of Membrane Science 2014;469:80–87.

[81] Sheath P, Majumder M. Flux accentuation and improved rejection in graphene-based filtration membranes produced by capillary-force-assisted self-assembly. Philosophical Transactions of the Royal Society A. 2016;374(2060):20150028.

[82] Han Y, Xu Z, Gao C. Ultrathin graphene nanofiltration membrane for water purification. Advanced Functional Materials. 2013;23(29):3693–3700.

[83] Prince JA, Bhuvana S, Anbharasi V, Ayyanar N, Boodhoo KVK, Singh G. Ultra-wetting graphene-based membrane. Journal of Membrane Science 2016;500:76–85.

[84] Hu M, Mi B. Enabling graphene oxide nanosheets as water separation membranes. Environmental Science & Technology. 2013;47(8):3715–3723.

[85] Yin J, Zhu G, Deng B. Graphene oxide (GO) enhanced polyamide (PA) thin-film nanocomposite (TFN) membrane for water purification. Desalination 2016;379:93–101.

[86] Sun P, Zhu M, Wang K, Zhong M, Wei J, Wu D, et al. Selective ion penetration of graphene oxide membranes. ACS Nano. 2013;7(1):428–437.

[87] Sun P, Zheng F, Zhu M, Song Z, Wang K, Zhong M, et al. Selective trans-membrane transport of alkali and alkaline earth cations through graphene oxide membranes based on cation–π interactions. ACS Nano. 2014;8(1):850–859.

[88] Joshi RK, Carbone P, Wang FC, Kravets VG, Su Y, Grigorieva IV, et al. Precise and ultrafast molecular sieving through graphene oxide membranes. Science. 2014;343(6172):752–754.

[89] You Y, Sahajwalla V, Yoshimura M, Joshi RK. Graphene and graphene oxide for desalination. Nanoscale. 2016;8(1):117–119.

[90] Liu B, Wu R, Baimova JA, Wu H, Law AW-K, Dmitriev SV, et al. Molecular dynamics study of pressure-driven water transport through graphene bilayers. Physical Chemistry Chemical Physics. 2016;18(3):1886–1896.

[91] Sun P, Wang K, Wei J, Zhong M, Wu D, Zhu H. Effective recovery of acids from iron-based electrolytes using graphene oxide membrane filters. Journal of Materials Chemistry A. 2014;2(21):7734.

[92] Tang YP, Paul DR, Chung TS. Free-standing graphene oxide thin films assembled by a pressurized ultrafiltration method for dehydration of ethanol. Journal of Membrane Science 2014;458:199–208.

[93] Lin L, Liu Z. Graphene synthesis: on-the-spot growth. Nature Materials. 2016;15(1):9–10.

[94] Wu T, Zhang X, Yuan Q, Xue J, Lu G, Liu Z, et al. Fast growth of inch-sized single-crystalline graphene from a controlled single nucleus on Cu–Ni alloys. Nature Materials. 2015;15(1):43–47.

[95] Liu Y, Xie B, Zhang Z, Zheng Q, Xu Z. Mechanical properties of graphene papers. Journal of the Mechanics and Physics of Solids. 2012;60(4):591–605.

[96] Hu W, Peng C, Luo W, Lv M, Li X, Li D, et al. Graphene-based antibacterial paper. ACS Nano. 2010;4(7):4317–4323.

[97] Cui Y, Kundalwal SI, Kumar S. Gas barrier performance of graphene/polymer nanocomposites. Carbon 2016;98:313–333.

[98] Zhu Y, Murali S, Cai W, Li X, Suk JW, Potts JR, et al. Graphene and graphene oxide: synthesis, properties, and applications. Advanced Materials. 2010;22(35):3906–3924.

[99] Robati D, Mirza B, Rajabi M, Moradi O, Tyagi I, Agarwal S, et al. Removal of hazardous dyes-BR 12 and methyl orange using graphene oxide as an adsorbent from aqueous phase. Chemical Engineering Journal 2016;284:687–697.

Chapter 3

Laser-Scribing Technology for Wafer-Scale Graphene Devices

He Tian, Mohammad Ali Mohammad, Wen-Tian Mi,
Yi Yang and Tian-Ling Ren

Additional information is available at the end of the chapter

http://dx.doi.org/10.5772/64228

Abstract

Graphene has attracted a lot of attention due to its amazing properties. A huge number of novel devices, covering the electric, acoustic, photonic, magnetic and mechanical domains, can be developed with graphene. Its ultrahigh mobility can enable ultra-fast transistors or photodetectors. However, the natural zero bandgap of graphene, with insufficient on/off ratio, limits its practical applications. In this chapter, we introduce laser-scribing technology that enables wafer-scale production of graphene devices. Moreover, such laser-scribed graphene (LSG) is, infact, semi-reduced graphene oxide with a finite bandgap, which is suitable for practical applications. We show five kinds of representative LSG devices and their integration. These devices are a resistive memory, an earphone, a strain sensor, a pressure sensor and a light-emitting device. These LSG devices are high-performance, flexible and low cost, which demonstrates the practical nature of laser-scribed graphene-based materials. Finally, an outlook is presented regarding how laser scribing, a serial patterning method, may lead to similar developments in various other serial lithography techniques, such as ion beam lithography.

Keywords: graphene, graphene oxide, two-dimensional device, laser-scribing technology, laser reduction, nanoelectronics

1. Introduction

Graphene is a well-known two-dimensional material that has attracted wide attention due to its unique properties [1], which have not been observed in three-dimensional (bulk) materials. Graphene's mobility at room temperature can be as high as 200,000 $cm^2 V^{-1} s^{-1}$ [2], which can be regarded as within the realm of ballistic transport and is much higher than the mobility of silicon. Moreover, graphene can absorb light in a wide range, even up to the far infrared [3], which is much broader than the capability of a conventional photodetector. Another powerful property is that its Young's modulus is 1 TPa [4], which makes it one of the strongest materials ever known. Graphene's thermal conductivity is also amazingly high at 5000 W/mK [5], which is also a record. Based on graphene's superior properties, a huge number of novel devices can be made using graphene, covering the electric, acoustic, photonic, magnetic and mechanical domains. However, graphene is not perfect. Its natural zero bandgap, with an insufficient on/off ratio, limits its practical applications. Especially, in logic circuits, the transistor needs to be turned off with a low current, which can reduce its power. This cannot be done in a graphene-based transistor, since the on/off ratio is only less than 10. Moreover, the current cannot be saturated under a high drain bias, which is also a drawback.

Figure 1. GO film obtained via the vacuum filtration method. (a) GO film on filter paper. (b) GO film on glass. (c) GO film on polyethylene terephthalate (PET). Reprinted by permission from Macmillan Publishers Ltd: Nature Nanotechnology (3:270), copyright (2008).[8].

A number of researchers tried to artificially introduce a bandgap in graphene. There are mainly three ways to do this. The first two ways consist of using a bilayer graphene and a graphene nanoribbon. The bandgap of bilayer graphene can only be opened around 30 meV, which is still not large enough. The graphene nanoribbon can provide sufficient on/off ratio, when the width of the nanoribbon is reduced down to 10 nm. However, the fabrication of graphene nanoribbon down to 10 nm is very challenging. Since these two methods cannot be scaled up, we refrain from discussing the details here. In this chapter, we mainly focus on the third method—using reduced graphene oxide (rGO). The bandgap of rGO can be tuned from 0~1.9 eV by controlling the quantity or level of reduction from graphene oxide (GO). Large scale graphene films have been prepared by chemical methods at low cost. A graphene oxide (GO) dispersion could be obtained by the Hummer's method [6], and the graphene oxide film can be prepared through various methods [7], that is spin-coating, spray-coating and vacuum filtration. Ultimately, the graphene film could be obtained by reduction. Some traditional methods include thermal and chemical reduction. Since there would be some oxygen functional groups remaining in the film after reduction, the width of the bandgap depends on how much oxygen will remain in the film. This can be tuned by controlling, for example, the reduction temperature. Large scale GO films can be prepared by a vacuum filtration method. This technique was firstly reported by Eda et al. (**Figure 1**), which revealed the application potential of rGO in transparent electrodes and transistors [8]. The next step is how to control the shape of rGO.

Although large scale graphene films could be obtained by various methods, the laser-scribing procedure [9, 10] was developed to achieve a precise pattern of graphene, which had never been attempted before. The graphene pattern can be simply and cost-effectively obtained by a single-step laser scribing, even on flexible substrates.

The laser-scribing technology is a method in which a laser is used to convert the graphene oxide into reduced graphene oxide. With the use of the LightScribe DVD drive, specific patterns can be obtained. The laser (wavelength 788 nm, power 5 mW) in the LightScribe DVD drive can reduce the GO to rGO (**Figure 2a**). The laser can be controlled by a software, such as the Nero StartSmart software, and thus the input image can be printed onto the GO as rGO. An in-plane transistor array design is shown in **Figure 2b**, which could be revealed on the graphene film in ~25 min by laser scribing. The golden film in **Figure 2c** is graphene oxide and the black part is multilayer graphene. Complex graphics could also be converted into graphene patterns, such as the Tsinghua university logo, shown in **Figure 2d** and **e**. And a coloured picture could be converted into a greyscale graphene pattern (**Figure 2f** and **g**), in which the greyscale directly corresponds to the degree of reduction.

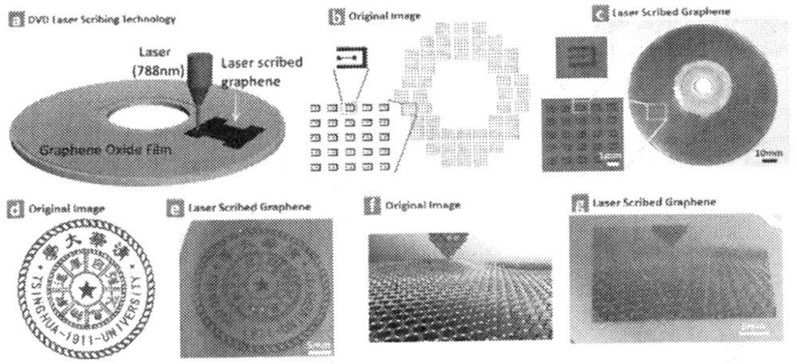

Figure 2. Demos of laser-scribed graphene. (a) Laser-scribing platform. (b) Layout for wafer-scale in-plane transistors. (c) Printed wafer-scale in-plane transistors. (d) An original image of the Tsinghua University logo. (e) The reproduced Tsinghua logo using LSG technology. (f) An original colourful image. (g) The printed LSG pattern in grey scale. Reprinted by permission from Macmillan Publishers Ltd: Scientific Reports (4:3598), copyright (2014) [16].

2. Characterization of laser-scribed graphene

In this section, we will demonstrate how to produce laser-scribed graphene(LSG) and discuss its characterization. The spot size of the laser, which is 20 μm in this work, determines the precision of the laser-scribed graphene. Post-exposure, it could be observed under an optical microscope that the laser-scribed graphene consists of long micro-ribbons (**Figure 3**). **Figure 3a** and **b** are the optical images of the laser-scribed graphene micro-ribbons under low magnification. Under visible-light irradiation, the GO is white and the rGO is black.

Increasing the magnification, a white line in the centre of the rGO micro-ribbon could be observed. Remarkably, different focus planes of GO and rGO, which are clearly shown in **Figure 3c** and **d**, suggest that the laser-scribed graphene micro-ribbon has a 3D structure. The total length of the rGO micro-ribbon can be very long (up to cm scale). If roll-to-roll technology is integrated with such laser-scribing technology, the length of rGO micro-ribbons can be even longer up to the meter scale. The most striking feature of the laser-scribing technology is that the rGO patterns can be obtained in a single step, which is time-efficient and low cost.

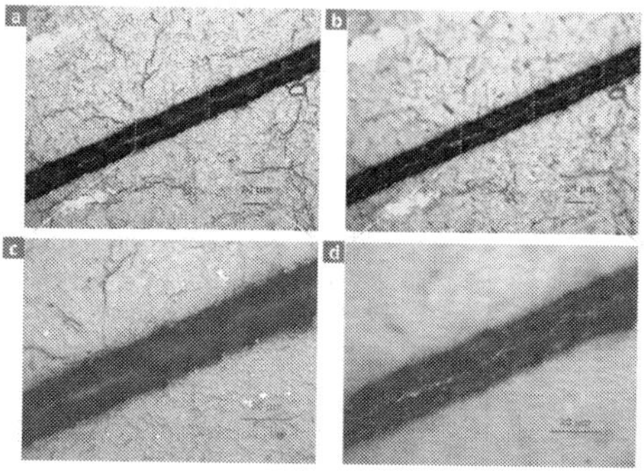

Figure 3. The optical image of a laser-scribed graphene micro-ribbon. (a) An optical image with focus on the GO surface. (b) The same optical image with focus on the rGO surface. (c) The zoomed-in image showing: (a) the GO surface and (d) the rGO surface, with z-direction profile.

Figure 4. A false coloured SEM image of a laser-scribed graphene micro-ribbon. The red part shows the GO surface, and the yellow part shows the LSG foam-like structure. Reprinted from Ref. [14] (CC BY 4.0).

A falsely coloured SEM image (**Figure 4**) of laser-scribed graphene reveals the 3D structure more clearly. The red part in the image represents the GO, and the yellow part represents the

http://dx.doi.org/10.5772/64228

rGO. The contrast of light and shade in the image of the rGO ribbon indicates the valley in the centre and the elevation on the two sides, which resemble the foam-like structure.

The structural change of graphene before and after the laser scribing is shown in **Figure 5**. The thickness of the laser-scribed graphene is 10 μm, which is 10 times larger than that of the original GO film. It could be observed, in the magnified image, that the rGO is comprised of loosely stacked graphene layers and the GO film is made of dense layers. In the case of thermal reduction, a totally opposite behaviour is observed. The thickness of the graphene film decreases after thermal reduction, which results from the loss of the oxygen functional groups. However, the high-energy laser beam rapidly converts the oxygen functional groups into a gas containing oxygen, which finally leads to the loose structure of rGO and an increase of the total thickness.

Figure 5. The lateral view of the laser-scribed graphene 3D structure. (a) A SEM profile showing the LSG and GO. (b) Zoomed-in image showing the loosely stacked graphene layers. (c) Zoomed-in image showing the dense GO film. Reprinted by permission from Macmillan Publishers Ltd: Scientific Reports (4:3598), copyright (2014) [16].

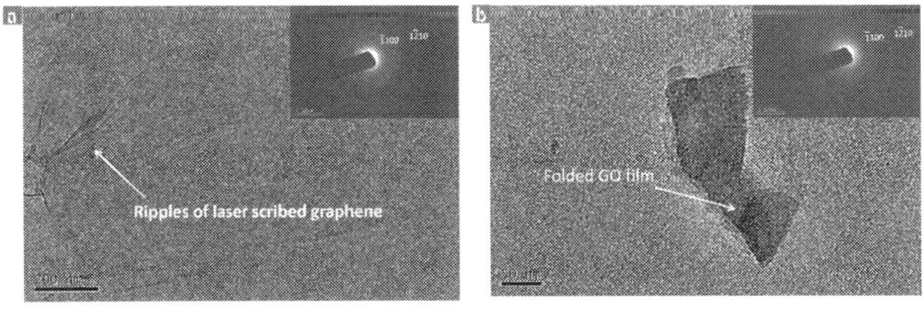

Figure 6. TEM images of (a) laser-scribed graphene and (b) GO (insets in both panels show the diffraction patterns from the atomic structure). Reprinted by permission from Macmillan Publishers Ltd: Scientific Reports (4:3598), copyright (2014) [16].

The diffraction pattern of the LSG (**Figure 6a**) confirms the hexagonal cellular lattice of graphene. On the contrary, the absence of a crystal lattice in the diffraction pattern of the stacked GO film (**Figure 6b**) indicates the broken C-C covalent bond, which results from the lattice defects bonding with oxygen functional groups.

Raman characterization is a quite efficient method to identify the properties of graphene. There are mainly three peaks present in graphene's Raman spectrum: the D-band (1350 cm^{-1}), G-band (1580 cm^{-1}) and 2D-band (2700 cm^{-1}). The D-band indicates the defects of the carbon lattice. If the D-band is strong, defects are present. The G-band position is sensitive to the doping level of graphene. If there is a right shift of the G-band, it means that the graphene is doped. The 2D-band/G-band ratio shows the layer number information. If the ratio is larger than 1, the film under inspection should be monolayer graphene. If it is around 1, then this is a strong indication of bilayer graphene. When the layer thickness number is larger than 3, the ratio is lower than 1. As shown in **Figure 7**, the change of the lattice structure could also be revealed in the Raman spectrum. The decrease of the D peak intensity suggests that part of the C–C bonds were repaired after the laser scribing. However, since there are still some defects left in the laser-scribed graphene, the D peak remains obvious. Besides, the left shift of the G peak indicates that the doping level of graphene decreased after the reduction, resulting from the removal of oxygen. And the appearance of the obvious 2D peak after the laser scribing indicates that the stacked multilayer graphene is generated after reduction.

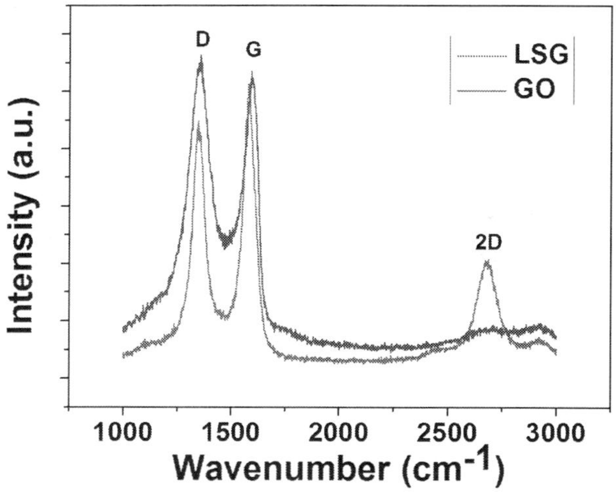

Figure 7. Raman spectra of GO and rGO.

X-ray photoelectron spectroscopy (XPS) is commonly used to identify the functional groups of various materials. Through the binding energy, the chemical formula and the electronic state of the elements can be revealed. The component of the laser-scribed graphene is shown in the XPS spectra (**Figure 8**). As shown in **Figure 8a**, the upper red spectrum is from the LSG and the lower black spectrum is from the GO. Comparing the LSG spectrum with the initial GO film spectrum, the oxygen peak obviously decreases for LSG. This is a direct evidence that the laser can reduce the oxygen from GO. The C1s peaks are analysed carefully to identify the functional groups on the carbon lattice. After fitting the sub-peaks, the GO film contains the C–C, disorder, carbonyl and π–π bonds. By comparing the LSG, it shows that the peaks of the C–C sp² bond and π-π bond are enhanced after laser reduction.

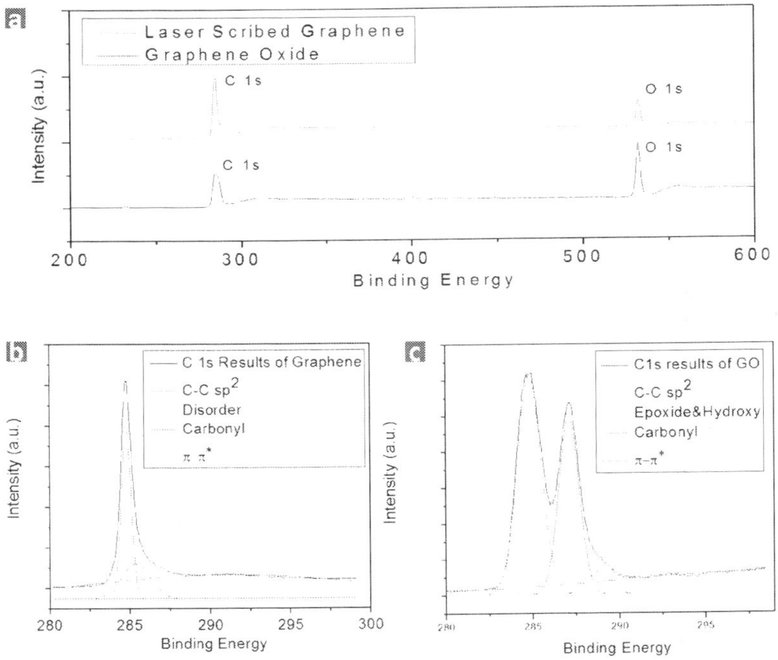

Figure 8. XPS spectra of GO and rGO. (a) The whole spectra of GO and rGO with the C_{1s} and O_{1s} peaks. (b) C1s peak of rGO. (c) C1s peak of GO.

3. LSG devices and their application

In this section, five different devices are presented, namely a memory device, an earphone, a strain sensor, a pressure sensor and a light emitter; all fabricated using laser-scribed graphene patterning technology. Each device will be presented in brief, while the readers are referred to the relevant citations to explore further details. Finally, multiple types of devices are patterned on the same surface in a single step, in close proximity, demonstrating the tremendously useful integration potential of this technique.

3.1. Graphene resistive random access memory

Using laser-scribing technology, a flexible graphene resistive random access memory or resistive memory (RRAM) [11] with a Fin-FET-like structure (**Figure 9a**) can be developed, which has great potential to increase the robustness of the device.

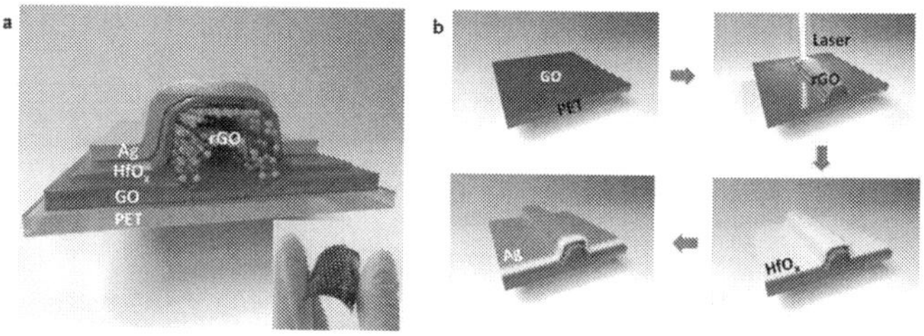

Figure 9. Flexible graphene RRAM device structure and fabrication process. (a) Device structure. (b) Fabrication process. Reprinted with permission from Nano Letters 14:3214. Copyright (2014) American Chemical Society [11].

As shown in **Figure 9b**, the GO dispersion was first spin-coated on the flexible polyethylene terephthalate (PET) substrates, following which the GO could be converted into rGO at specific locations. Then, the HfO_x and the silver top electrodes were deposited and patterned consecutively. For further details of the fabrication process and the testing method, please refer to Ref. [11].

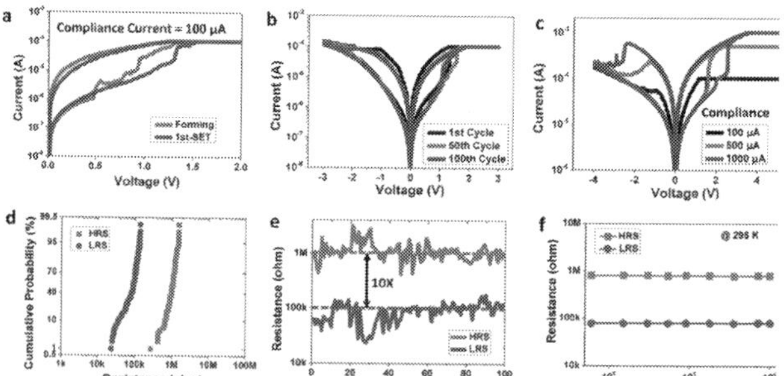

Figure 10. Measurement results of the flexible graphene RRAM. (a) Forming process. (b) Switching performance in the first, 50th and 100th cycles. (c) Switching curves under different compliance currents. (d) High- and low-resistance-state distribution. (e) High- and low-resistance-state under different cycle numbers. (f) Retention testing up to 10^4 s for both high- and low-resistance-state. Reprinted with permission from Nano Letters 14:3214. Copyright (2014) American Chemical Society [11].

The excellent performance of the flexible graphene RRAM is clearly shown in **Figure 10**, including the forming-free feature (**Figure 10a**), repeatability (**Figure 10b**), multi-valued feature (**Figure 10c**), uniformity (**Figure 10d**), feasible memory window (**Figure 10e**) and stability (**Figure 10f**).

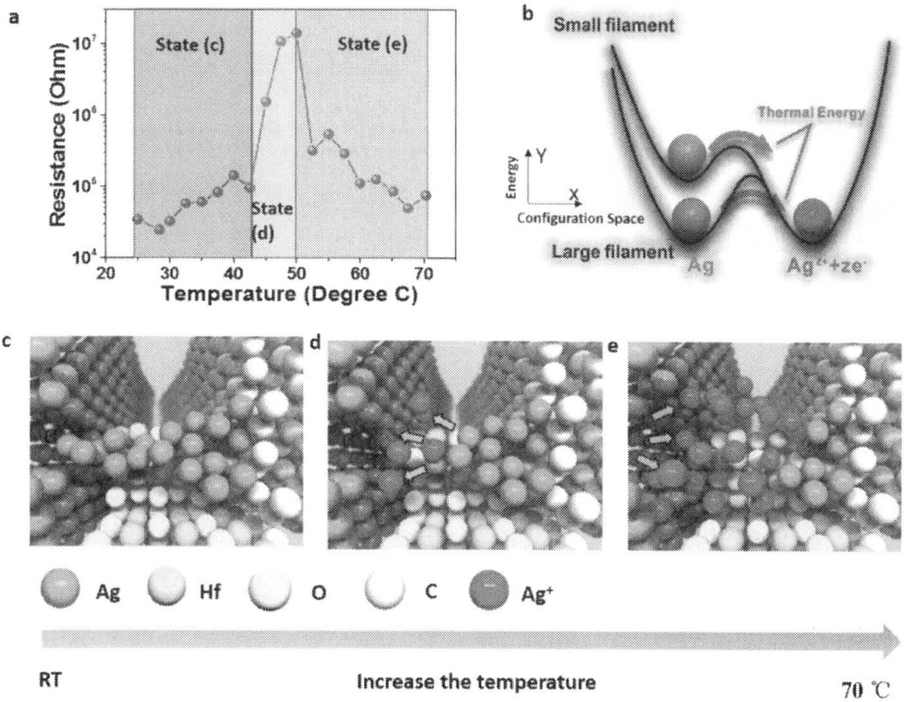

Figure 11. The working principle of the LSG RRAM. (a) The on-resistance *versus* the temperature. (b) The energy diagram for the filament formation. (c,d,e) The dynamic process in the filament region under increasing temperatures. Reprinted with permission from Nano Letters 14:3214. Copyright (2014) American Chemical Society [11].

The working principle of the graphene RRAM could be revealed by the temperature characteristics analysis (**Figure 11**). When the temperature is raised, the increase of the resistance indicates that the conductive filaments are made of silver. As the temperature is continually raised, some silver filaments transform into silver ions, leading to the increase of the resistance. As the silver ions are also conductive, if they increase, the resistance falls again.

This working principle may be confirmed by electrical measurements. The threshold voltage is only 0.5 V when the top electrode is made of silver. But for the device with Pt top electrodes, the threshold voltage is 2 V, which suggests that the soft breakdown voltage of the HfO_x is 2 V. Therefore, it is obvious that the HfO_x could not be broken down under 0.5 V, to form the oxygen filament, which confirms the claim that the filament is made of silver.

3.2. Graphene earphone

Flexible graphene earphones[12] can also be fabricated based on the laser-scribing technology. Compared with traditional earphones, the graphene earphone has the characteristics of flexibility and ultra-low thickness.

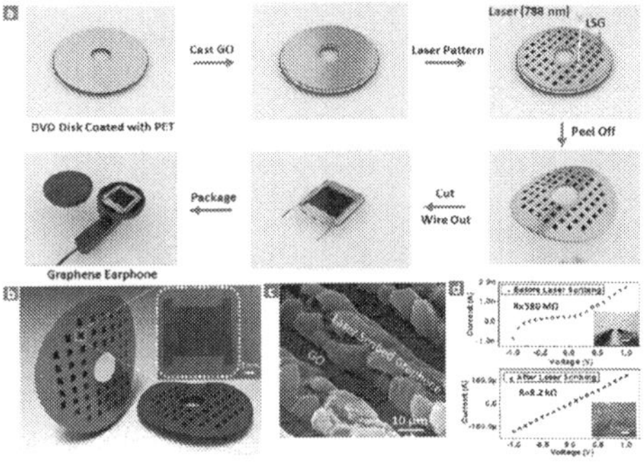

Figure 12. Graphene earphone. (a) Fabrication process of the graphene earphone using laser-scribing technology. (b) Wafer-scale graphene earphone after the laser scribing (inset showing the image of a single device). (c) SEM image showing the LSG profile. (d) I–V electrical properties before and after the laser scribing. Reprinted with permission from ACS Nano 8:5883. Copyright (2014) American Chemical Society [12].

Using the laser-scribing technology, the graphene sound source array could be fabricated on a PET substrate (**Figure 12a** and **b**). The I–V curve indicates the good conductivity of the graphene device. The sound source was wired out and packaged into a commercial earphone case (**Figure 13**). Further details about the fabrication and testing can be found in Ref. [12].

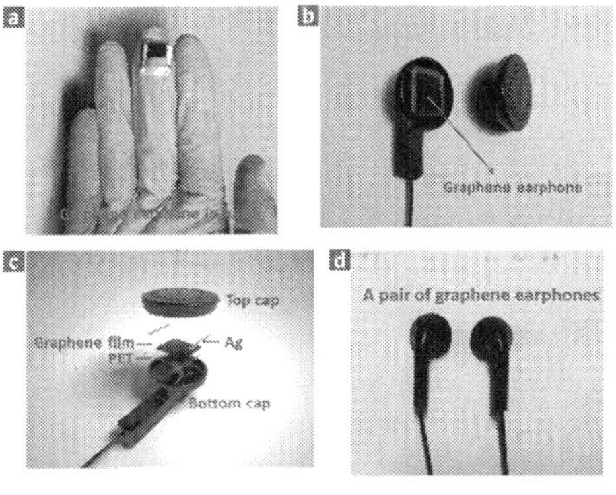

Figure 13. Packaged graphene earphone. (a) Graphene earphone in hand. (b) Graphene earphone after packaging. (c) The 3D assembled structure of the graphene earphone. (d) A pair of graphene earphones. Reprinted with permission from ACS Nano 8:5883. Copyright (2014) American Chemical Society [12].

The acoustic properties of the graphene earphone are tested on the measurement system, as shown in **Figure 14a**. The test results indicate that the sound pressure level of the frequency spectrum is flatter than that of commercial earphones, especially in the ultrasonic band. Moreover, the frequency spectrum of the graphene earphone covers not only the audible band (from 20 Hz to 20 kHz for human being) but also the ultrasonic band (from 20 to 50 kHz), which is the audibility zone for some animals. This implies that the graphene earphone could be a useful tool for interspecies communication, which is also demonstrated in **Figure 15**. The 35 kHz acoustic signal is conveyed to a dog via graphene earphones, and with the aid of some prior training, the dog stood up as soon as the signal was delivered (**Figure 15**).

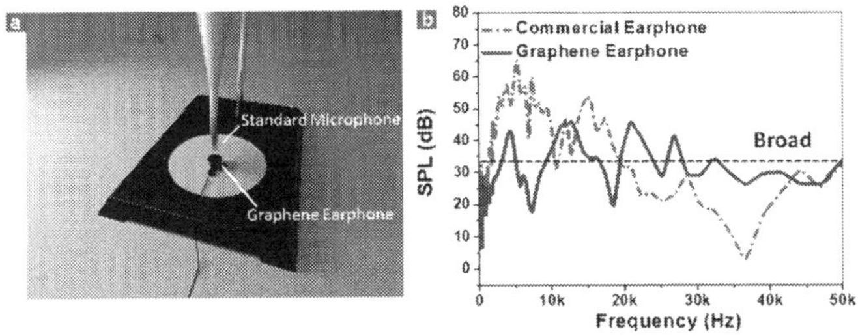

Figure 14. Performance testing of the graphene earphone. (a) Acoustic test platform for the graphene earphone. (b) Acoustic spectrum of the graphene earphone compared with a commercial one. Reprinted with permission from ACS Nano 8:5883. Copyright (2014) American Chemical Society [12].

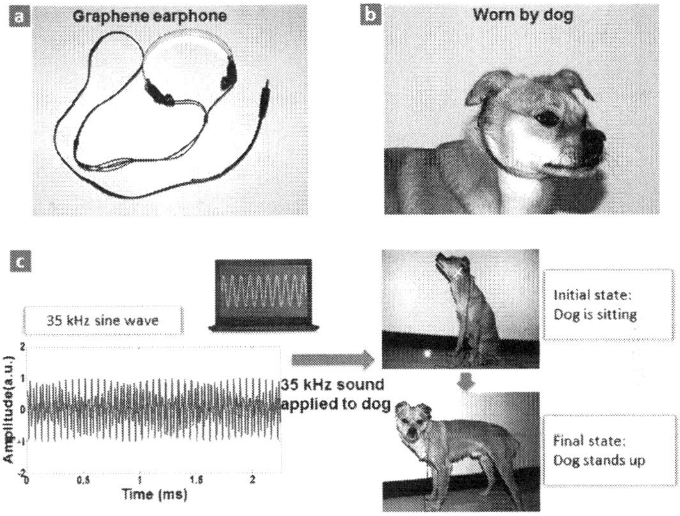

Figure 15. The graphene earphone as a tool for interspecies communication. (a) A pair of graphene earphones. (b) A dog wearing a graphene earphone. (c) Sound waves at 35 kHz applied to the dog signals it to stand up. Reprinted with permission from ACS Nano 8:5883. Copyright (2014) American Chemical Society [12].

3.3. Graphene strain sensor

Strain sensors are widely used in various application areas, but most of them are based on rigid substrates, which limit the sensitivity of the sensor. Using laser-scribing technology, a flexible graphene strain sensor [13] is achieved with greatly enhanced sensitivity. The sensitivity (GR) can be expressed by the following equation:

$$\mathrm{GF} = (\Delta R / R) / \varepsilon \tag{1}$$

Where $\Delta R/R$ is the relative resistance change, and ε is the applied strain.

The pattern of the graphene strain sensor array could be fabricated on the GO film via laser-scribing technology on a flexible PET substrate with low cost and high speed (**Figure 16**). The laser-scribed graphene is made of a loose stack of graphene layers, which could greatly enhance the sensitivity of the graphene strain sensor.

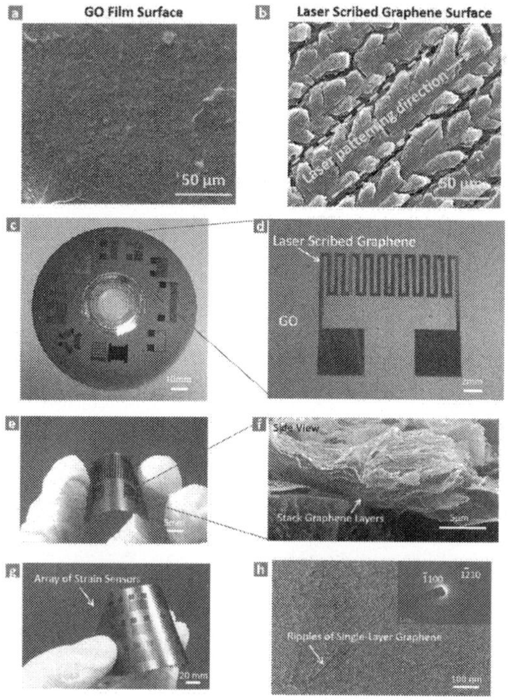

Figure 16. Graphene strain sensor. (a) SEM image of the GO surface. (b) LSG surface showing the direction of the laser movement. (c) Wafer-scale strain sensor with different shapes. (d) One strain sensor image. (e) Strain sensor showing good flexibility. (f) Cross-section image of the LSG film. (g) Array of graphene strain sensors. (h) TEM image showing the single-layer graphene. Reprinted from Ref. [13] with permission of The Royal Society of Chemistry.

The test results of the graphene strain sensor are shown in **Figure 17**. The sheet resistance was recorded while stretching the sensor. The sensitivity calculated from the data is 9.49, and **Figure 17f** indicates the rapid response of the graphene strain sensor. Commercial strain

http://dx.doi.org/10.5772/64228

sensors based on metal have a gauge factor of 2~5. Our LSG strain sensor has a much higher sensitivity as compared to commercially available strain sensors, which enables it to have a wide variety of applications. The working principle can be explained as follows: LSG contains a lot of graphene sheets. The conductivity of LSG depends on the overlapping area of the graphene sheets. When the LSG is stretched, the overlapping area of graphene sheets decreases, which increases the resistance. Therefore, by measuring the resistance change, one can know the strain applied to the LSG.

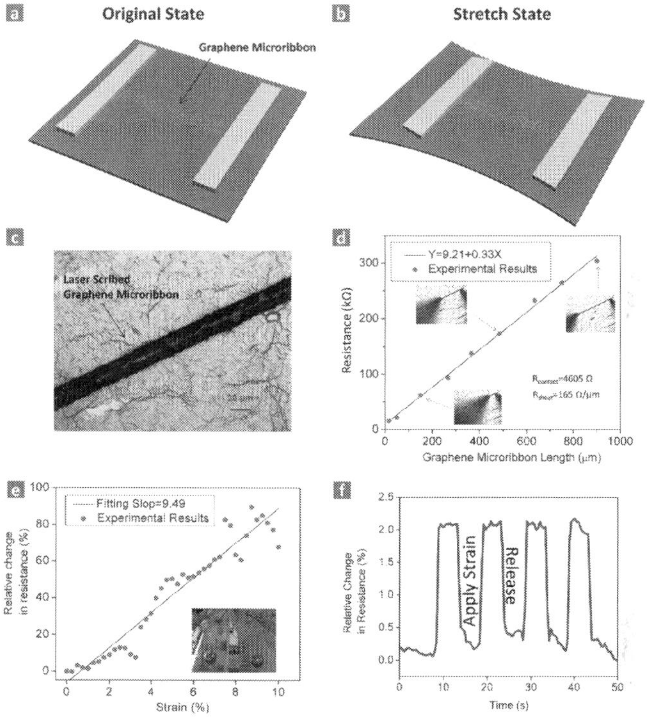

Figure 17. Test results of the graphene strain sensor. (a) Graphene micro-ribbon sensor in its original state. (b) Graphene micro-ribbon sensor in its stretched state. (c) Optical image of the graphene micro-ribbon. (d) Resistance *versus* length. (e) Relative resistance change *versus* strain. (f) Dynamic response to instantaneous force. Reproduced from Ref. [13] with permission of The Royal Society of Chemistry.

3.4. Graphene pressure sensor

Graphene pressure sensors can also be obtained by laser-scribing technology [14]. Conventional pressure sensors only have good sensitivity at pressure ranges lower than 5 Pa. However, the sensitivity will obviously decrease when the pressure is larger than 5 Pa. It is thus necessary to develop pressure sensors with high sensitivity over a wide range of pressures.

As shown in **Figure 18**, the core device structure is a stacking over of the two LSG layers with the patterns oriented perpendicularly to one another. The upper layer is oriented in the *y*-direction, and the lower layer is oriented in the *x*-direction. The profile of the LSG has a "V" shape with a height of 10.7 μm and a width of 19.8 μm. The fabrication process is illustrated in **Figure 19**, and the complete process details are provided in Ref. [14].

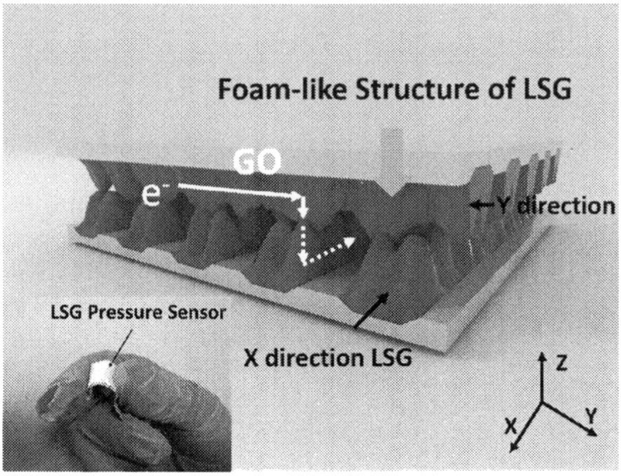

Figure 18. Device structure of LSG pressure sensor (the inset shows the real device in hand). Reprinted from Ref. [14] (CC BY 4.0).

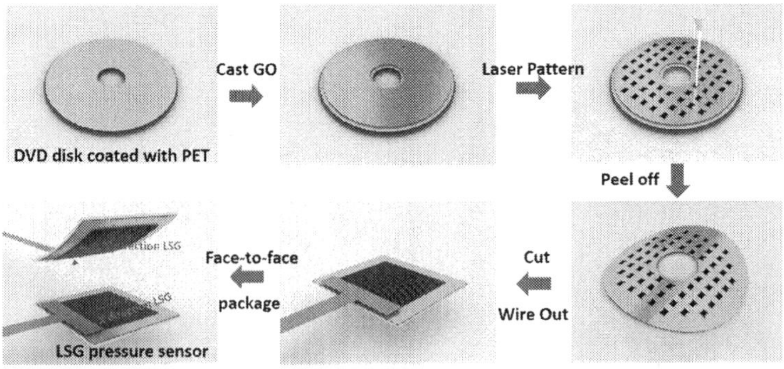

Figure 19. Fabrication process for the LSG pressure sensor. Reprinted from Ref. [14] (CC BY 4.0).

The working principle of the LSG pressure sensor can be described as follows (**Figure 20**): when a force is applied on top, the inter layer distance of the LSG is reduced, which can change the resistance of the LSG. Since LSG has a porous structure, the force can not only increase the contact area but also increase the density of the LSG. Both these effects can increase the current pathway. When the pressure is released, the LSG foam releases and the interlayer distance recovers, which can lower the current to its initial level.

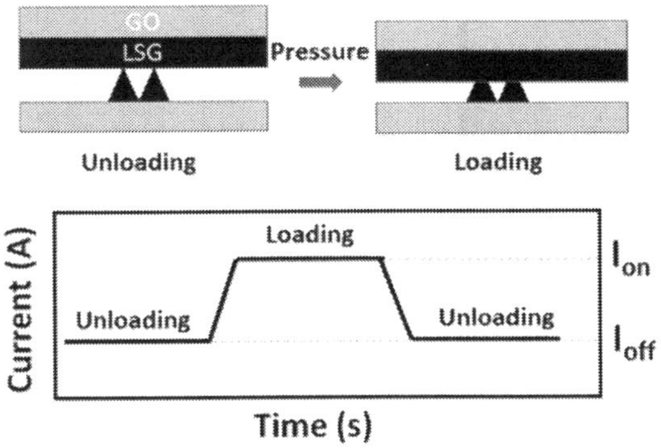

Figure 20. The working principle of the LSG pressure sensor. Reprinted from Ref. [14] (CC BY 4.0).

In order to test the LSG pressure sensor, we have set up a force testing system. This system contains a motor, a pressure sensor and a resistance analyser. In this system, a static force up to 113 kPa can be applied and the resistance analyser can acquire the resistance from the LSG pressure sensor simultaneously. As shown in **Figure 21**, if the pressure sensing range is up to 50 kPa, the sensitivity can be as high as 0.96 kPa^{-1}. This indicates that the contact area and density obviously change in this pressure range. When the pressure is larger than 50 kPa, the pressure sensor starts to saturate with a sensitivity of 0.005 kPa^{-1}. This demonstrates that we have successfully obtained a high sensitivity pressure sensor using laser-scribing technology.

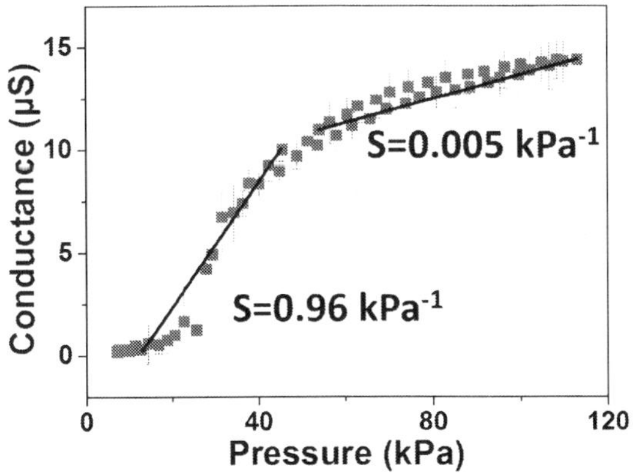

Figure 21. Testing results of the LSG pressure sensor. Reprinted from Ref. [14] (CC BY 4.0).

3.5. Graphene light-emitting device

In conventional light-emitting devices, the emission wavelength cannot be changed after the devices are fabricated. This is because the bandgap of the material cannot be modified. Previously, there was no report on light-emitting devices with tunable wavelength in a wide range. Here, we obtained a wavelength tunable graphene light-emitting device based on laser-scribing technology [15].

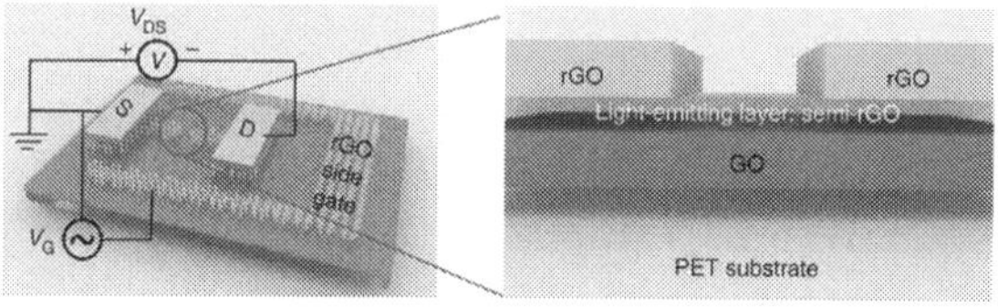

Figure 22. Device structure of the graphene light-emitting device. The semi-rGO layer functions as the light-emitting layer. Reprinted from Ref. [15] (CC BY 4.0).

As shown in **Figure 22**, a graphene in-plane transistor can be fabricated, based on laser-scribing technology. An in-plane transistor means that the gate, source and drain are all in the same plane. The gate electrical field can be applied to channel through the in-plane GO dielectric. All the components, namely the channel, source, drain and gate can be fabricated in a single laser-scribing step. After the device fabrication, a large current is applied to the channel (mA), which can burn the upper rGO away and expose the interface layer consisting of semi-rGO. A striking feature is that the whole fabrication process does not need vacuum or high temperature.

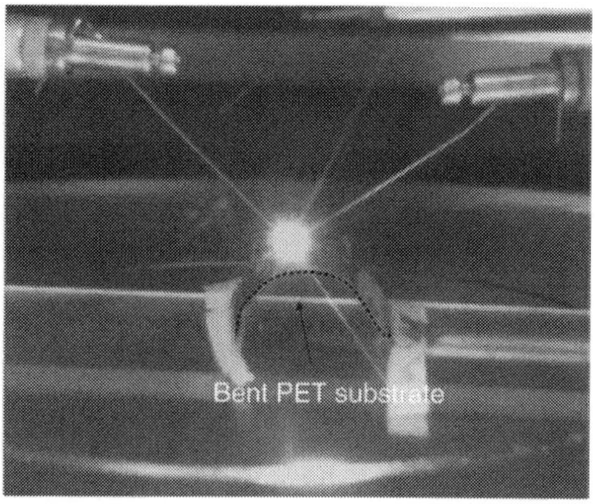

Figure 23. Flexible graphene light-emitting device with red light emission. Reprinted from Ref. [15] (CC BY 4.0).

http://dx.doi.org/10.5772/64228

When the device is driven by a voltage of 10 V, there is a noticeable red light emission (**Figure 23**). Moreover, as it is directly fabricated on a PET substrate, the whole device is flexible. **Figure 24** shows the light emission spectra under different gate voltages. Each curve shows a single Lorentzian shape. When the gate voltage is zero, the light wavelength is around 690 nm. When the gate voltage is 50 V, the wavelength shifts to 470 nm. This indicates that increasing the gate voltage can shift the wavelength to a lower value. Moreover, the blue light emission efficiency is much lower than the red light efficiency from the intensity shown in the spectrum. Under different gate voltages, the selected semi-rGO is excited to emit light. The light emission mechanism can be explained by the Poole–Frenkel effect. Under a strong electrical field, the electrons can be excited by impact ionization effect and recombined with the holes. The extra energy can be released by emission of photons.

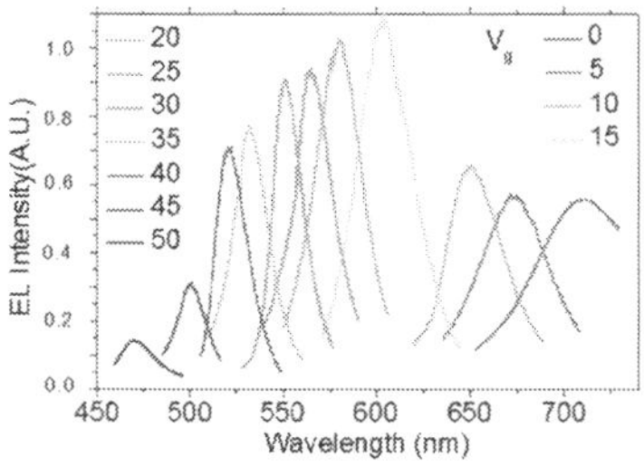

Figure 24. Different light emission spectra, obtained by tuning the gate voltage. Reprinted from Ref. [15] (CC BY 4.0).

3.6. Wafer-scale graphene devices integration

Since a variety of graphene devices can be easily patterned onto the GO film via the laser-scribing technology, it is possible to integrate multifunctional graphene devices [16] over an entire wafer (**Figure 25**). As shown in **Figure 25a**, a graphene transistor, photodetector and loudspeaker were integrated on the same chip. Furthermore, the substrates could even be flexible (**Figure 25b**). The patterning process of a whole wafer was finished in 25 min. This development brings our field one-step closer to the concept of heterogeneous integration, and it would be useful to investigate this technique and platform in further detail.

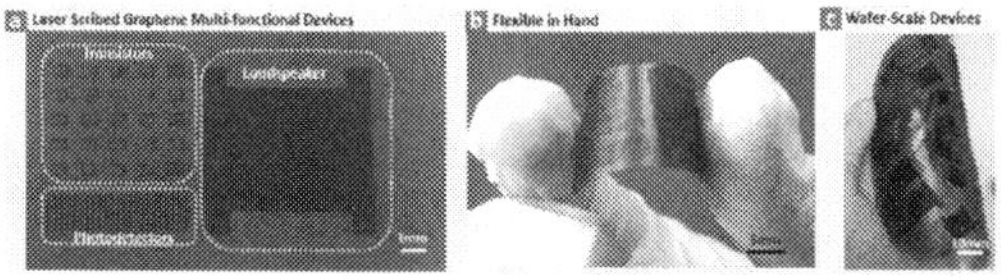

Figure 25. Wafer-scale graphene devices integration. (a) Multifunctional graphene devices containing transistors, photodetectors and loudspeakers. (b) Flexible graphene devices. (c) Wafer-scale graphene devices. Reprinted by permission from Macmillan Publishers Ltd: Scientific Reports (4:3598), copyright (2014) [16].

4. Outlook: ion scribing for graphene oxide reduction

The outstanding success of laser scribing for rGO-based device fabrication naturally leads one to imagine whether other serial writing techniques can be used as well. Besides laser lithography, other serial writing techniques include electron beam lithography, ion beam lithography, proton beam lithography and X-ray (or synchrotron) lithography. A review of literature reveals that reduction of GO has already been briefly investigated with electron beam lithography [17] and ion beam lithography [18]. Although the conversion efficiency of electron beam lithography for graphene oxide reduction is very low, and hence impractical, ion beam lithography is very promising. Lobo et al. [18] have investigated the morphology, composition and conductivity changes and confirmed the reduction of GO by Gallium ions. We have also begun investigations using ion beam lithography for GO reduction and device fabrication, and we observed that the conductivity of 300 nm GO films on SiO_2/Si conducts up to 1.8×10^{-10} A (0.18 pA) for a Ga^+ ion dose of 750 μC/cm², as shown in **Figure 26** [19].

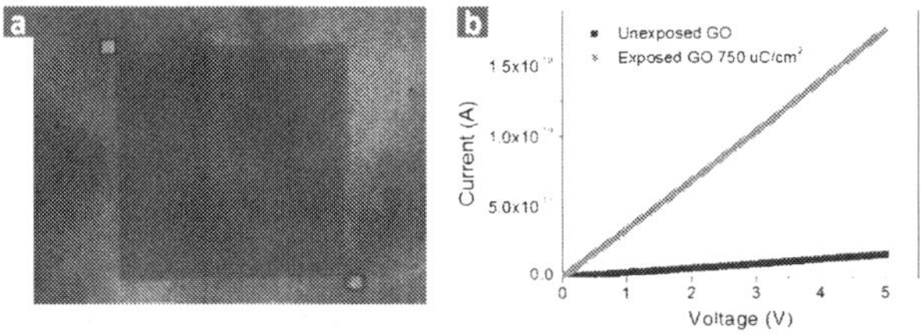

Figure 26. GO reduction by Ga^+ ion beams showing (a) the exposed and unexposed areas, and (b) the increase in conductivity at a dose of 750 μC/cm², as compared with unexposed GO.

5. Summary

In this chapter, the fabrication of wafer-scale graphene devices by one-step laser-scribing technology is demonstrated. Five kinds of novel graphene devices were developed, including a memory, an earphone, a strain sensor, a pressure sensor and a light-emitting device. The graphene resistive memory has a Fin-like structure with forming-free behaviour, stable switching, reasonable reliability and potential for 2-bit storage. The graphene earphone enables wide-band sound generation from 100 Hz to 50 kHz, which can be used for both humans and animals. The strain sensor based on graphene micro-ribbons has a gauge factor up to 9.49. The sensitivity of the graphene pressure sensor is as high as 0.96 kPa^{-1} in a wide pressure range (0~50 kPa). A graphene light-emitting device is also developed with a tuneable emission wavelength. These results demonstrate that laser-scribing technology can be used as a platform to develop novel graphene devices. As a serial lithography technique, at the forefront of controlled graphene oxide reduction, laser scribing serves as a model for similar developments in other patterning techniques, such as ion beam lithography.

Author details

He Tian[1,2], Mohammad Ali Mohammad[1,3], Wen-Tian Mi[1], Yi Yang[1] and Tian-Ling Ren[1*]

*Address all correspondence to: RenTL@tsinghua.edu.cn

1 Institute of Microelectronics and Tsinghua National Laboratory for Information Science and Technology (TNList), Tsinghua University, Beijing, China

2 Ming Hsieh Department of Electrical Engineering, University of Southern California, Los Angeles, CA, United States

3 School of Chemical and Materials Engineering (SCME), National University of Sciences and Technology (NUST), Sector H-12, Islamabad, Pakistan

References

[1] Geim AK, Novoselov KS. The rise of graphene. Nature Materials. 2007;6:183-191. DOI: 10.1038/nmat1849

[2] Bolotin KI, Sikes KJ, Jiang Z, Klima M, Fudenberg G, Hone J, Kim P, Stormer HL. Ultrahigh electron mobility in suspended graphene. Solid State Communications. 2008;146(9):351-355. DOI:10.1016/j.ssc.2008.02.024

[3] Mak KF, Ju L, Wang F, Heinz TF. Optical spectroscopy of graphene: From the far infrared to the ultraviolet. Solid State Communications. 2012;152(15):1341–1349. DOI: 10.1016/j.ssc.2012.04.064

[4] Lee C, Wei X, Kysar JW, Hone J.. Measurement of the elastic properties and intrinsic strength of monolayer graphene. Science. 2008;321(5887):385-388. DOI: 10.1126/science.1157996

[5] Balandin AA, Ghosh S, Bao W, Calizo I, Teweldebrhan D, Miao F, Lau CN. Superior thermal conductivity of single-layer graphene. Nano Letters. 2008;8(3):902-907. DOI: 10.1021/nl0731872

[6] Hummers Jr, WS., Offeman RE. Preparation of graphitic oxide. Journal of the American Chemical Society. 1958;80(6):1339. DOI: 10.1021/ja01539a017

[7] Dreyer DR, Park S, Bielawski CW, Ruoff RS. The chemistry of graphene oxide. Chemical Society Reviews. 2010;39(1):228-240. DOI: 10.1039/B917103G

[8] Eda G, Fanchini G, Chhowalla M. Large-area ultrathin films of reduced graphene oxide as a transparent and flexible electronic material. Nature Nanotechnology. 2008;3(5): 270-274. DOI: 10.1038/nnano.2008.83

[9] El-Kady MF, Strong V, Dubin S, Kaner RB. Laser scribing of high-performance and flexible graphene-based electrochemical capacitors. Science. 2012;335(6074):1326-1330. DOI: 10.1126/science.1216744

[10] El-Kady MF, Kaner RB. Scalable fabrication of high-power graphene micro-super capacitors for flexible and on-chip energy storage. Nature Communications. 2013;12(4): 1475. DOI: 10.1038/ncomms2446

[11] Tian H, Chen HY, Ren TL, Li C, Xue QT, Mohammad MA, Wu C, Yang Y, Wong HS. Cost-effective, transfer-free, flexible resistive random access memory using laser-scribed reduced graphene oxide patterning technology. Nano Letters. 2014;14(6):3214–3219. DOI: 10.1021/nl5005916

[12] Tian H, Li C, Mohammad MA, Cui YL, Mi WT, Yang Y, Xie D, Ren TL. Graphene earphones: entertainment for both humans and animals. ACS Nano. 2014;8(6): 5883-5890. DOI: 10.1021/nn5009353

[13] Tian H, Shu Y, Cui YL, Mi WT, Yang Y, Xie D, Ren TL. Scalable fabrication of high-performance and flexible graphene strain sensors. Nanoscale. 2014;6(2):699-705. DOI: 10.1039/C3NR04521H

[14] Tian H, Shu Y, Wang XF, Mohammad MA, Bie Z, Xie QY, Li C, Mi WT, Yang Y, Ren TL. A graphene-based resistive pressure sensor with record-high sensitivity in a wide pressure range. Scientific Reports. 2015;5:8603. DOI: 10.1038/srep08603

[15] Wang X, Tian H, Mohammad MA, Li C, Wu C, Yang Y, Ren TL. A spectrally tunable all-graphene-based flexible field-effect light-emitting device. Nature Communications. 2015;6:7767. DOI: 10.1038/ncomms8767

[16] Tian H, Yang Y, Xie D, Cui YL, Mi WT, Zhang Y, Ren TL. Wafer-scale integration of graphene-based electronic, optoelectronic and electroacoustic devices. Scientific Reports. 2014;4:3598. DOI: 10.1038/srep03598

[17] Chen L, Xu Z, Li J, Min C, Liu L, Song X, Chen G, Meng X. Reduction and disorder in graphene oxide induced by electron-beam irradiation, Materials Letters. 2011;65:1229. DOI: 10.1016/j.matlet.2011.01.063

[18] Lobo DE, Fu J, Gengenbach T, Majumder M. Localized deoxygenation and direct patterning of graphene oxide films by focused ion beams. Langmuir. 2012;28:14815. DOI: 10.1021/la303369m

[19] Mohammad MA, Mi WT, Li YX, Tian H, Yang Y, Ren TL. Manuscript in preparation (2016).

Chapter 4

Light Emission from Graphene

Young Duck Kim and Myung-Ho Bae

Additional information is available at the end of the chapter

http://dx.doi.org/10.5772/64051

Abstract

Graphene has attracted great interest due to its superior and unique physical properties. The strong light-matter interaction with ultrafast and broadband photoresponse of graphene has allowed for the development of state-of-the art optical components, such as photodetectors and optical modulators, making this material very promising for ultrafast optical communications. However, so far, light emission from graphene has remained elusive, although this would open the door towards obtaining atomically thin, flexible and transparent light sources and graphene-based on-chip interconnects. In this chapter, we review experimental results and techniques of the electrically driven light emission from graphene in the infrared and visible spectrum range.

Keywords: Graphene, light emission, power dissipation, high electric field, Phonon temperature

1. Introduction

In order to obtain bright thermal radiation from gapless graphene, the radiative electron–hole recombination process is not enough due to the ultrafast energy relaxation that occurs through electron–electron (~10 fs) and electron–phonon (~1 ps) interactions. On the other hand, graphene's superior mechanical strength, high-temperature stability and non-equilibrium electron temperature, compared to lattice temperature, may enable the efficient thermal radiation under high electric field. However, the thermal radiation from electrically biased graphene is significantly influenced by the substrate. Especially, dominant vertical heat dissipation through underlying substrate, and extrinsic scattering effects, such as charged impurities and the surface polar optical phonon, limit the thermal radiation efficiency of graphene.

Freely suspended structure enables the dramatic suppression of undesirable vertical heat dissipation and extrinsic scattering effects by substrate, and much more promising to yield efficient and brighter thermal radiation from graphene in the infrared to visible range. Furthermore, the emitted light from suspended graphene interacts with the reflected light from the separated substrate surface, resulting in interference effects that can be used to tune the radiation spectrum of this material. Electro-thermal engineering will allow obtaining advanced graphene light emitters for atomically thin, flexible and transparent lighting and ultrafast optical interconnects. In this chapter, we will provide an overview of the thermal radiation from electrically biased graphenes in the infrared and visible spectrum range.

2. High electric field transport in graphene on substrate

Graphene's superior electronic [1], mechanical [2] and thermal [3] properties can be used for high mobility field-effect transistors (FETs) [4] and mechanical resonators [5]. In the case of electronic devices, as shown in **Figure 1a**, that depicts a graphene FET, the applied electric field applied by the source-drain voltage generates a Joule self-heating effect, which leads to a change of its electronic transport behaviour or device failure, in some cases [6]. For carbon nanotubes (CNTs), the observed electronic transport characteristics have been explained by heat transports along the CNT channels and between CNTs and the substrates [7]. There have been reports on the heat dissipation from graphene devise in a diffusive regime, where the chemical potential changes along the graphene channel. A direct method to study the heat dissipation is to accurately measure the temperature during device operation (see Sections 2.1–2.3). The temperature measurement of graphene on a substrate biased by a source–drain voltage has been performed in four ways: (1) shift of Raman peaks [8–11]; (2) thermal scanning imaging [11]; (3) thermal radiation [9, 10, 12] and (4) ratio of the Stoke/anti-Stoke intensities in Raman spectroscopy [9, 10, 13]. Methods (1) and (2) are sensitive to the acoustic phonon (AP) temperature (T_{ap}), while methods (3) and (4) are linked to the electron and optical phonon (OP) temperatures (T_e and T_{op}), respectively.

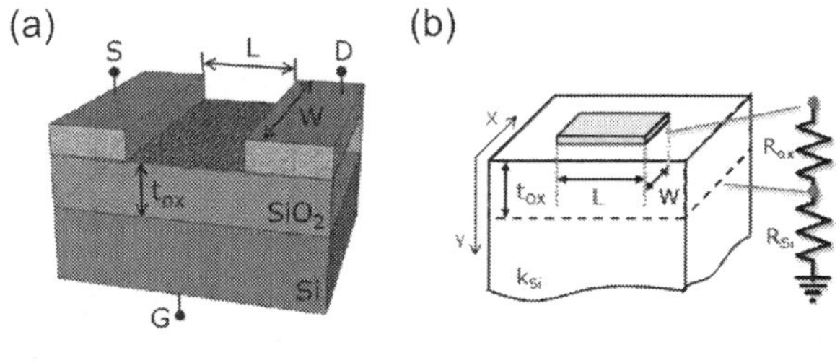

Figure 1. (a) Graphene field-effect transistor. S, D and G are source, drain and gate, respectively. L and W are length and width of graphene channel, respectively, and t_{ox} is thickness of SiO_2 layer [14]. (b) Thermal model for a graphene on SiO_2/Si substrate. R_{ox} and R_{Si} are thermal resistance of the oxide and silicon substrate, respectively. K_{si} is thermal conductivity of Si substrate.

The Fourier's law for thermal conductance is:

$$\dot{Q}=G_{th}\Delta T, \tag{1}$$

where $\dot{Q}$ is the amount of heat transferred per unit time, ΔT is the temperature difference between two points of interest, and G_{th} is the thermal conductance. Here, the temperature corresponds to T_{ap} since the acoustic phonons are the main contributors to heat conductions. **Figure 1b** shows how to apply the Fourier's law to graphene devices on a substrate, where R_{ox} and R_{Si} are the thermal resistances of the SiO_2 layer and Si substrate, respectively. The spreading thermal resistance of the Si substrate is given by $R_{Si} \sim 1/[2k_{si}(LW)^{1/2}]$ and $R_{ox} \sim t_{ox}/(k_{ox}LW)$, where $k_{ox} = 1.3\ Wm^{-1}K^{-1}$ and $k_{si} = 50\ Wm^{-1}k^{-1}$ are the thermal conductivities of the SiO_2 layer and Si substrate at room temperature, respectively [12]. t_{ox} is the thickness of SiO_2 layer, and L (W) is the length (width) of the graphene channel. Eq. (1) can also be expressed as $\Delta T = R_{th} P$, where R_{th} is the total thermal resistance, and P is the electrical power.

Figure 2 shows a summary of temperature increase (ΔT) of graphene devices on substrates, as a function of electrical power density, determined by various temperature measurement techniques. The closed, opened and half-closed scattered points show the acoustic, T_{ap}, T_{op} and T_e and electron temperatures, respectively.

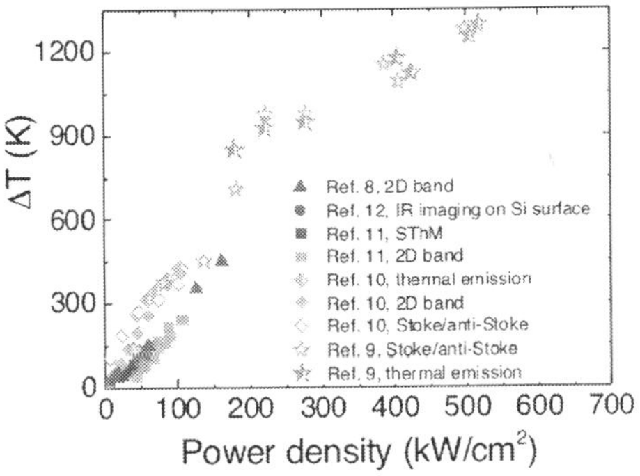

Figure 2. averaged temperature increase (ΔT) as a function of applied power per unit area of reported devices. Detailed information of each device is described in **Table 1**.

The detailed measurement techniques can be found in the legend of **Figure 2**. For each sample, one can obtain the R_{th} of the system, through the slope data plotted in **Figure 2**. **Table 1** and **Figure 3** show the R_{th} comparison between the experiments and calculations, based on the relation: $R_{th} \approx R_{ox} + R_{Si}$ (see **Figures 1b** and **3**).

References	L [μm]	W [μm]	t_{ox} [nm]	R_{ox} [K/W]	R_{Si} [K/W]	R_{cal} [K/W]	R_{exp} [K/W]
[8]	2.65	1.45	300	60,000	5100	65,100	73,500
[12]	28	6	300	1370	770	2140	2100
[11]	7	4	300	8240	1900	10,130	8300
[10]	4.15	1.45	300	38,350	4070	42,430	68,100
[9]	3.6	1.6	280	37,400	4160	41,560	69,800

Table 1. Summary of reported the thermal resistance of SiO_2 layer (R_{ox}), and Si substrate (R_{si}) with calculated (R_{cal}) and experimentally measured (R_{exp}) total thermal resistance, where all samples are placed on SiO_2/Si substrates.

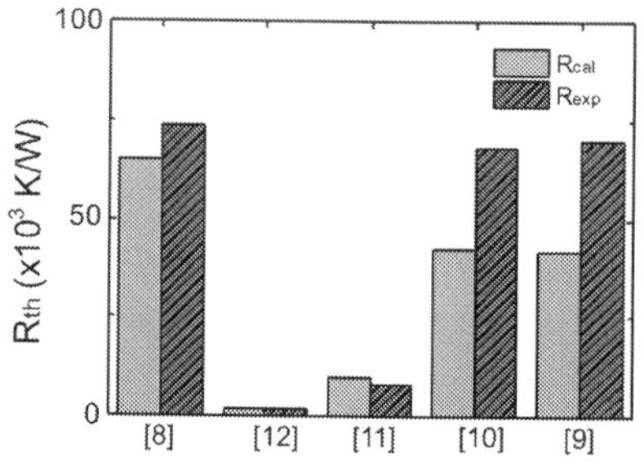

Figure 3. Average calculated (R_{cal}) and measured (R_{exp}) total thermal resistance of each reference, where R_{cal} and R_{exp} were adopted from **Table 1** for corresponding references.

In order to treat the heat transfer from graphene on a substrate into the surrounding environment, as shown in **Figure 1b**, let us first consider a one-dimensional heat equation along the graphene channel [12]:

$$\frac{d^2T}{dx^2} + p = \frac{g}{k_g A}(T - T_0), \quad (2)$$

where g is the thermal conductance of the substrate per unit length, k_g is the thermal conductivity of graphene, A is the cross-section of graphene, $p = IF_x$, I is the current and F_x is the electric field at x. For simplicity, if we consider an infinitely long graphene channel without bias current, Eq. (2) has a solution, $\Theta = \Theta(0)e^{-x/L_H}$, where $\Theta(x) = T(x) - T_0$ and $L_H = \sqrt{k_g A/g}$. The L_H is the lateral temperature diffusion (healing) length from a heat source through the graphene. The heat in a graphene device on a substrate is dissipated into a heat sink ($T = T_0$) mainly in two means: a direct way through the substrate underneath graphene or indirect way via metal contacts. In a graphene device where L is comparable to L_H, the heat dissipation through the metal contacts becomes important. In the opposite case of $L >> L_H$, the heat flow through the substrate is the dominant heat dissipation path. g is approximately given by $1/[L(R_{ox}+R_{Si})]$ (also see **Figure 1b**). For a monolayer graphene device on 300 nm thick, SiO_2/Si substrate with L =

http://dx.doi.org/10.5772/64051

25 μm, $W = 6$ μm and graphene thickness $t_g = 0.35$ nm, we get $L_H \approx \sqrt{\frac{t_{ox} t_g k_g}{k_{ox}}} \approx 0.2$ μm [12]. Here, we used the thermal conductivity of a few μm size graphene on a substrate as $k_g \sim 600$ $Wm^{-1}K^{-1}$ [15]. Thus, in general, graphene devices of few micrometres in length directly dissipate their heat through the substrate underneath graphene, and the temperature profile of the graphene reflects the heat generation profile.

2.1. Electron temperature of graphene

Thermal radiation can be observed for graphene devices on a substrate under high electric field, which is attributed to the electrically driven spontaneous emission from hot electrons of graphene [9, 10]. The thermalized charge carriers under electric field emit electromagnetic radiation, according to Planck's law, as a grey body:

$$I(E, T_e) = \varepsilon(E) \frac{2E^3}{h^2 C^2} \left[\exp\left(\frac{E}{k_B T_e} \right) - 1 \right]^{-1}, \tag{3}$$

where $I(E, T_e)$ is the spectral radiance, $\varepsilon(E)$ is the emissivity of graphene, E is the energy of the emitted photon, h is Planck's constant, c is the speed of light, and k_B is the Boltzmann constant.

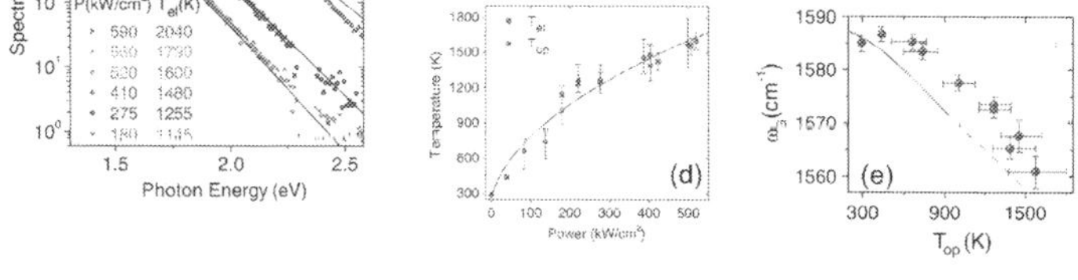

Figure 4. (a) Thermal radiation spectrum from graphene on substrate under applied electrical power per unit area (symbols). The radiation spectra are well fitted to Planck's law (solid line) and electron temperatures are estimated. (b) Anti-stoke and (c) Stokes Raman G mode as function of applied electric field and optical phonon temperature are indicated. (d) Electron and optical phonon temperature as function of the dissipated electrical power. It indicates the equilibrium of electron and optical phonon temperature. (e) Raman G mode frequency (symbol) features as function of optical phonon temperature. Solid line is the theoretically predicted temperature dependences under phonon equilibrium. Dashed line is linearly extrapolation (from Ref. [9]).

The T_e, extracted from its thermal radiation spectrum in the range of 1.3–2.6 eV, was of 1000–2000 K [9], as shown in **Figure 4a**. From the extracted T_e, we can assume that the strong electron-

electron interaction rapidly thermalizes electrons and holes, and attains hot charge carriers by the applied electric field. Furthermore, the experimentally measured emissivity $\varepsilon(E)$ of graphene of $=1.6 \pm 0.8$ % is consistent with the measured absorptivity $\alpha = 2.3$ % for monolayer graphene [9, 10]. This means that, for graphene under high electric field, Kirchhoff's law is valid in the measured energy range.

2.2. Optical phonon temperature of graphene

Raman spectroscopy is a well-known technique for the characterization of a number of layers, doping level, mechanical strain and temperature of graphene. In particular, for graphene devices under high electric field, Raman spectroscopy accurately provides the phonon temperature and population. In typical Raman spectroscopy, spontaneous Raman effects are due to the inelastic light scattering, with optical phonon modes of the material being determined by measurement. The resulting inelastic scattered of excited photon emits or absorbs optical phonons. The Stoke process involves the creation of extra optical phonons (n+1), and the anti-Stoke process is the annihilation of optical phonons (n).

In graphene, the phonon population for zone centre optical phonons can be estimated from the G peak signal intensity of the Raman spectra, which is proportional to phonon populations with effective temperature, and follows from the Bose-Einstein distribution. Raman signal intensity ratio of the anti-Stoke (I_{AS}) and Stokes (I_S) process is as follows:

$$\frac{I_{AS}}{I_S} \approx \exp\left(-\frac{\hbar\omega_G}{k_B T_{op}}\right), \tag{4}$$

where $\hbar\omega_G$ is the energy of G phonon. Experimentally, T_{op} is extracted from the ratio I_{AS}/I_S as shown in **Figure 4b** and **c**. Due to the high optical phonon energy of graphene, anti-Stoke Raman signals can be observed only above 440 K [9, 10, 13].

2.3. Acoustic phonon temperature of graphene

The Raman spectrum of graphene also exhibits a downshift of the Raman peak position with increasing temperature. This is due to the anharmonic coupling effect and anharmonic temperature of secondary phonons into G mode phonon decay and a thermal expansion effect [16]. Empirically, the down shift of the G peak can be measured as a linear function of T_{ap} as:

$$\omega_G = \omega_G^0 + \chi T_{ap}, \tag{5}$$

where ω_G^0 is the G peak frequency at zero temperature, and χ is the coefficient for temperature correction term of -0.016 cm^{-1} K^{-1} [17], at moderate temperature.

2.4. Non-equilibrium phonon population in graphene

As discussed in Sections 2.1–2.3, accurate measurements of graphene temperature provide a deep understanding of the heat dissipation mechanism and electron-phonon interaction in biased graphene devices. Direct comparison of T_e and T_{op} temperatures of electrically biased graphene on a substrate shows that the T_e and T_{op} are in equilibrium up to 2000 K, as shown in **Figure 4d**.

In the case of CNTs, non-equilibrium phonon distributions have been measured when electrically biased, as high-energy optical phonons have larger populations than low-energy phonons, such as radial-breathing mode and acoustic phonons [7, 18].

Similarly to what was found with CNTs, non-equilibrium phonon populations in electrically biased graphene on a substrate have been observed. **Table 1** and **Figure 3** [9, 10] showed that the elevated T_{op} could be estimated by the ratio of Stoke/anti-Stoke intensity of G phonons and T_e measured by the thermal emission. For both cases, the experimentally obtained thermal resistance, R_{exp} is much larger than the predicted thermal resistance, R_{cal}, which could be due to the optical phonons and electrons not being in equilibrium with the acoustic phonon. This is consistent with the observation of the temperature deviation between T_{op} and the anharmonic temperature of secondary phonons, for which G phonons decay for $T_{op} > 670$ K [9]. From the device case in Ref. [10], the anharmonic temperature extracted from the 2D band showed similar temperatures to the T_{op} and T_e up to $T_{op} \sim 720$ K. In the reports, however, we should note that the temperature calibration for the 2D band was only obtained for $83\,K < T < 373\,K$, while the measurements was performed at $300\,K < T < 700\,K$, thus temperature correction could be needed for the high temperature region.

Experimental results show that strong electron-optical phonon coupling enables the equilibrium of the electron and optical phonon temperature, whereas an energy relaxation bottleneck to low-energy phonons results in the non-equilibrium phonon population. Furthermore, strong coupling with surface polar optical phonons and substrate acoustic phonons provides an extra electronic energy relaxation path, resulting in the effective cooling of acoustic phonons of graphene, under high electric field. Therefore, non-equilibrium optical phonon temperature (T_{op}) is given by:

$$T_{op} = T_{ap} + \alpha(T_{ap} - T_0), \tag{6}$$

where T_0 is the ambient temperature environment, and α is the non-equilibrium coefficient. The α is defined by [7]:

$$\alpha = \frac{R_{op}}{R_{ap}}, \tag{7}$$

where R_{op} and R_{ap} are the thermal resistances related to optical phonon decay into acoustic phonons, and the heat conduction of acoustic phonons into the heat sink, respectively. Thus, the T_{ap} and T_{op} are given by $T_{ap} = T_0 + PR_{ap}$ and $T_{op} = T_{ap} + PR_{op}$, respectively. For the devices from reference [9, 10], R_{exp} was ~70,000 K/W while R_{cal} was ~40,000 K/W (see **Table 1** and **Figure 3**). Since R_{cal} corresponds to R_{ap} and $R_{exp} \sim R_{ap} + R_{op}$, we get $R_{op} \sim$ 30,000 K/W, resulting in $\alpha \sim 0.75$.

3. Infrared light emission from graphene on substrate

Graphene has unique electronics and phonon states under high electric field, such as decoupled electrons and low-energy phonon temperature [9, 13]. These make graphene an efficient material as thermal radiation source, determined only by T_e. Furthermore, gapless graphene exhibits high performance for optoelectronics such as photodetectors, optical modulator and plasmonic devices in the infrared range. Developing a high-speed and efficient infrared light source from graphene has a great potential for optical communications, as well as medical applications.

Previous studies of infrared light emission from graphene were accomplished by graphene FETs on a substrate under high electric field [9, 10, 12]. Due to the self-heating effect of graphene under electrical bias, only a small fraction (<10^{-6}) of energy is converted into light and most of the energy is dissipated into the substrate and metal contact. The extracted T_e of graphene light sources was reported to be in the range of 700–2000 K [9, 10].

In graphene FETs, charge carrier density profiles are determined by applied electric field and gate voltage. According to the self-consistent electrical-thermal model of charge and heat

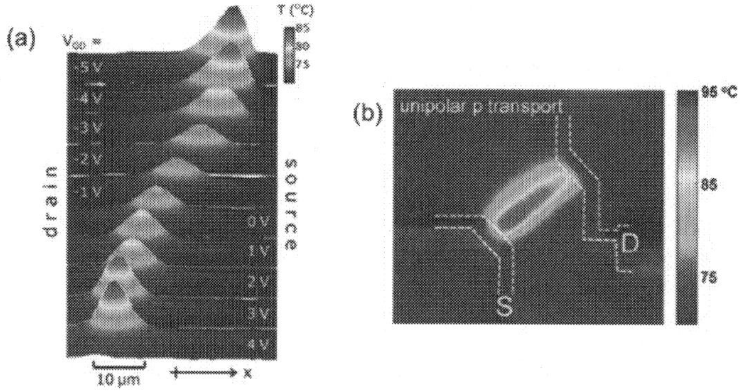

Figure 5. (a) Infrared mapping of temperature profiles along the graphene,showing hotspot formation. The hot spot moves from source to drain, marking the location of minimum charge density and maximum electric field, following the device electrostatics (from Ref. [14]). (b) Imaged temperature map of belayed graphene field-effect transistor (from Ref. [12])).

transport in graphene, the minimum charge carrier density spatial location is the cross-point of the Fermi level across the charge neutrality point. The hot spot position of the infrared image directly reveals the spatial location of the charge neutrality point along the graphene channel, as shown in **Figure 5a** and **b**. In some case, the stationary hot spot of infrared light emission, even under variation of gate voltages and source-drain bias, is attributed to the defects of graphene and trapped charges in the substrate [10, 12, 14].

Spontaneous thermal radiation is governed by the Fermi's golden rule, and the transition rate is determined by the optical density of the environment. This was shown in Ref. [19], through an optical micro cavity controlled graphene infrared light source, with emission around 925 nm,. Due to the strong light-matter interaction of atomically thin graphene and the optical confinement effect, micro-cavity structures enable the control of 20-fold enhancement of photocurrent generation, which is spectrally selective near infrared thermal radiation. The thermal radiation confinement effect inhibits spontaneous thermal radiation wavelengths larger than the resonant wavelength of the optical cavity. The optical cavity-induced suppression of the radiative thermal radiation also leads to the self-heating effect in graphene, which is affected by non-radiative heat transfer through the substrate, metal electrodes and radiative heat transfer.

4. High electric field transport in suspended graphene

As discussed in the previous section, high electric field transport of graphene on a substrate is affected by the substrate and environmental conditions. Limited thermal radiation efficiency and electron temperature are due to the dominant heat dissipation to the substrate and extrinsic scattering effects, such as electron-hole puddle, trapped charges [20] and surface polar phonon [21]. According to some authors [22, 23], suspended graphene shows ultrahigh mobility and fractional quantum Hall effect at low temperature, with low electric field. Therefore, in order to observe the intrinsic properties of graphene under high electric field, suspended structures are essential.

In suspended graphene, the self-heating effect plays a role due to significantly reduced heat dissipation through the substrate. Thermally induced charge carriers, saturation velocity, charge carrier mobility and thermal conductivity of graphene are significantly affected by the applied electric field and temperature.

In the clean limit of suspended graphene, under high electric field as shown in **Figure 6a–c**, the total charge carrier density of graphene is dominated by thermally induced charge carriers (n_{th}), represented as:

$$n_{th} = \left(\frac{\pi}{6}\right)\left(\frac{k_B T}{\hbar v_F}\right)^2, \qquad (8)$$

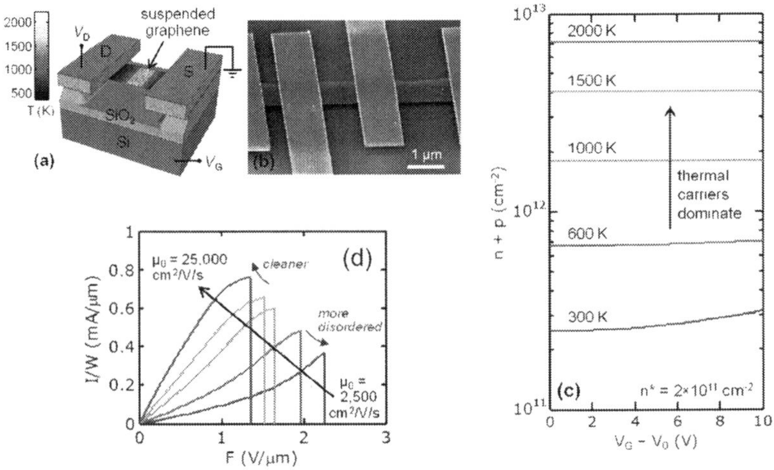

Figure 6. (a) Suspended graphene device, where colour scale indicates the temperature under high electric field. (b) SEM image of suspended graphene grown by CVD. (c) Calculated total charge carrier density as function of gate voltage with increasing temperature. In suspended graphene, thermally generated charge carriers are dominant at high temperature and independent of gate voltage. (d) Simulated I/W versus F with variation of low-field mobility-based electro-thermal self-consistent simulation (from Ref. [24]).

where v_F is the Fermi velocity of graphene. The intrinsic carrier density of graphene is $n_i=4(n_{th}^2+(n_{pd}/2)^2)$, where n_{pd} is electron-hole puddle-induced carriers. The carrier density induced by electrostatic gate V_g is $n_g=C_{ox}(V_{dr}-V_g)/e$, where V_{dr} is the charge neutral voltage of graphene. In total, the electron (hole) carrier density (see **Figure 6c**) is as follows:

$$n_e(n_h)=\frac{1}{2}\left(\pm n_g+\sqrt{n_g^2+n_i}\right). \tag{9}$$

The current density J in the suspended graphene, with an applied electric field, is presented by the continuity equation:

$$J=e(n_{ex}+n_{hx})v_{dx}, \tag{10}$$

where e is the electron charge, n_{ex} (n_{hx}) is the electron (hole) density, v_d is the drift velocity along the graphene channel. The drift velocity is expressed by:

$$v_{dx}=\frac{\mu_x F_x}{[1+(\mu_x F_x/v_{sat})^{\eta}]^{1/\eta}}, \tag{11}$$

where μ_x is the temperature dependent charge carrier mobility, F_x is the electric field along the graphene channel, and v_{sat} is a the saturation velocity determined by the intrinsic graphene optical phonon (160 meV) scattering [24, 25].

For a self-consistent electrical-thermal calculation of suspended graphene, the heat diffusion equation for the temperature profile (T_x) along the graphene channel takes the from:

$$\frac{d^2T_x}{dx^2} + \frac{1}{k_x Wt}\frac{dP_x}{dx} - \frac{2g_x}{k_x t}(T_x - T_0) = 0, \tag{12}$$

where P_x $(= I(V_x - IR_c))$ is the local dissipated electrical power, k_x is the temperature dependent thermal conductivity, W is width, t is thickness, g_x is the thermal conductance per unit area ($g_x \sim 0$ in vacuum), and T_0 is ambient temperature. Assuming that the temperature-dependent charge carrier mobility is considered, $(T_e) = \mu_0(T_0/T_e)^\beta$, where μ_0 is the ambient temperature mobility, T_e is the electron temperature, and β varies from 1.5 to 2.5, depending on the kind of sample. Furthermore, thermal conductivity is assumed to be $k(T_{ap}) = k_0(T_0/T_{ap})^\gamma$, where k_0 is the ambient temperature thermal conductivity, T_{ap} is the acoustic phonon temperature of graphene, and γ varies from 1.7 to 1.9 depending on the kind of sample.

Using a self-consistent electrical-thermal calculation in suspended graphene, as shown in **Figure 6d**, clean samples show the saturation behaviour under high electric field. This implies that high charge carrier mobility of suspended graphene devices have a larger temperature dependence ($\beta \sim 2.5$) due to the strong electron-intrinsic graphene optical phonon interactions. However, disordered suspended graphene shows superliner behaviour and less dependence on temperature ($\beta \sim$ 1.5). This is due to dominant hopping transport with increase of thermally generated charge carriers under high electric field [24].

In contrast with suspended CNTs, suspended graphene does not show a negative differential conductance, due to the linear increase of density of states in 2D, whereas CNTs have strong 1D phonon scattering, resulting in a negative differential conductance [7].

Along the temperature profile of suspended graphene, drift velocity and charge carrier density are affected by current density. Theoretically, saturation velocity of graphene was predicted as $v_{sat} \sim 2.7 \times 10^7 cm/s$, which is limited by 160 meV intrinsic graphene optical phonons [26]. Experimentally, saturation velocity before breakdown is measured as $v_{sat} \sim 1.7 \times 10^7 cm/s$ [24]. Of course, in a suspended graphene structure, flexural phonons [27] cannot be ignored, as they may be one of reasons for variation in saturation velocity.

Suspended graphene is well known as a material having the highest thermal conductivity at room temperature ($k_0 \sim$ 2500 W/mK). However, thermal conductivity of suspended graphene displays an abrupt decrease ($\sim T^{-1.7}$) at high temperature [24]. This is attributed to the flexural phonons modes which enable strong second-order three-phonon scattering [28] ($\sim T^{-2}$) and first-order Umklapp phonon–phonon scattering ($\sim T$). The estimated average thermal conductivity of electrically biased suspended graphene at 1000 K is of ~310 W/mK, for both mechanically exfoliated and chemical vapour deposited (CVD) grown graphene.

5. Bright visible light emission from suspended graphene

Graphene has a great potential as an efficient thermal radiation light source due to its superior mechanical strength, high-temperature stability, high current density and non-equilibrium phonon modes. However, previous reports regarding the thermal radiation of supported graphene [9, 10, 12, 19] have been limited to the near infrared light emission range, and extremely small fractions of electrical power are converted into the photons (~10^{-6}). The main reason for the limited performance of graphene light emitters supported on a substrate is the dominant heat dissipation through the substrate, and significant hot carriers cooling by dominant extrinsic scattering origins, such as charged impurities, trapped charges and surface polar phonons. In the case of freely suspended graphene, these issues can be ignored. Therefore, suspended structures are promising for the observation of bright, efficient light emission radiation approaching the intrinsic properties.

There are several methods that can be used for the synthesis of suspended graphene structures. One is wet-etching of a sacrificial layer after metallization onto the graphene. The other procedure is the transfer of patterned graphene onto pre-fabricated trench substrate. In order to increase the fabrication yield and prevent the collapse of graphene, a critical point dryer process is important. Suspended graphene structures are fabricated using mechanically

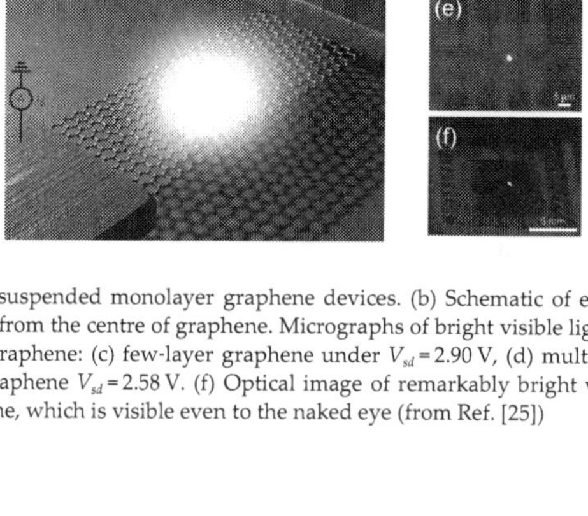

Figure 7. (a) SEM image of suspended monolayer graphene devices. (b) Schematic of electrically biased suspended graphene and light emission from the centre of graphene. Micrographs of bright visible light emission from a suspended mechanically exfoliated graphene: (c) few-layer graphene under V_{sd} = 2.90 V, (d) multi-layer graphene under V_{sd} = 7.90 V, and (e) monolayer graphene V_{sd} = 2.58 V. (f) Optical image of remarkably bright visible light emission from a suspended few-layer graphene, which is visible even to the naked eye (from Ref. [25])

exfoliated graphene or large-scale graphene grown by the CVD method. **Figure 7a** shows the obtained suspended graphene devices [25].

In order to observe light emission from suspended graphene under electrically bias, an experimental set-up, similar to that shown in **Figure 7b**, is used under vacuum ($\sim 10^{-5}$ Torr), at room temperature. In addition, to get clean graphene channel and reliable contact resistance, a slow current-induced annealing process [29] is essential. Suspended graphene exhibits bright visible light at the centre of the graphene channel once the applied bias voltage exceeds the critical field ($0.5 \sim 1.0$ V/μm), as shown in **Figure 7c–e**. The intensity of visible light is remarkably intense and can even be observed by the naked eye, without additional magnification, as shown in **Figure 7f**.

Y.D. Kim et al. [25] also demonstrated multiple visible light emission from a large-scale suspended graphene array, which was fabricated from large-scale CVD graphene. The synthesis of a large-scale, robust and bright visible light emitter array from CVD graphene will allow obtaining graphene-based flexible and transparent lighting and display modules. Furthermore, complementary metal-oxide-semiconductor (CMOS) compatibility of graphene technology will enable a hybrid graphene-photonic platform for ultrafast optical communications.

The radiation spectrum from electrically biased suspended graphene over a trench of depth ($D \sim 1000$ nm) exhibits multiple emission peaks in the visible range, as shown in **Figure 8a**. These multiple emission peaks are quite interesting, since graphene is a gapless material, and the radiation spectrum was predicted to follow the featureless Planck's law of a grey body (Eq. (3)). The multiple peaks in the visible range are strongly affected by the trench depth rather than by the number of layers of graphene and electronic band structure.

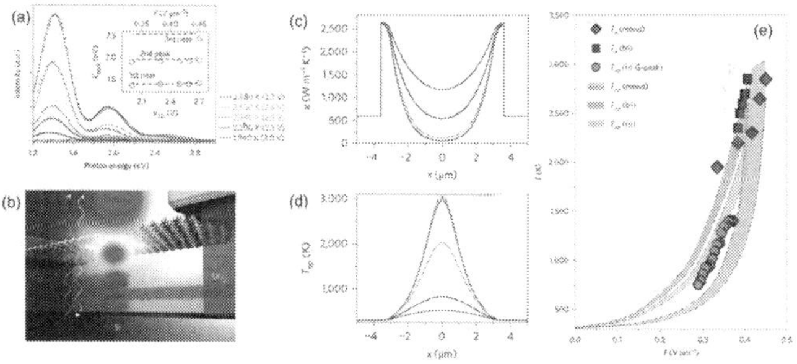

Figure 8. (a) Spectra of visible light emitted from electrically biased suspended graphene exhibiting multiple strong emission peaks. Thermal radiation with interference effect (solid line) fit well to the experimental data (symbol). Estimated electron temperatures of suspended graphene are indicated as function of applied source-drain bias. Inset: emission peak energies as a function of source-drain bias and applied electric field. (b) Interference effect between reflected (dashed arrow) and thermal radiation originated directly from graphene suspended over a trench. (c) Estimated thermal conductivity and (d) optical phonon temperature of suspended graphene based on self-consistent electrical and thermal transport, where it is assumed that $T_{op} = T_e$. (e) Measured temperature of graphene (symbol) and calculated temperature (dashed line) as function of applied electric field (from Ref. [25]).

These multiple emission peaks in the visible range and significant modulation by trench depth can be understood by interference effects between the light emitted directly from the suspended graphene and light reflected from the substrate as shown in **Figure 8b**. When we neglect the light reflection and absorption by the graphene, the interference effect on thermal radiation is given by:

$$I(\omega, D) = I_0(\omega)(\frac{1+|r(\omega)|}{2} + Re\left[r(\omega)\exp\left(\frac{i2\omega D}{c}\right)\right], \tag{13}$$

where $I_0(\omega) \sim \omega^3 / (\exp(\hbar\omega / k_B T_e) - 1))$ is the thermal radiation intensity of suspended graphene, $r(\omega)$ is the reflection coefficient of Si substrate, ω is the photon frequency, is the emissivity of graphene, D is the trench depth, k_B is the Boltzmann constant, T_e is the electron temperature of graphene, and c is the speed of light [25]. A strong interference effect enables the selectively enhancement of thermal radiation from graphene at a certain wavelength and can increase the radiation efficiency through the engineering of interference effects.

Based on thermal radiation with interference effect (Eq. 13), electron temperature of suspended graphene was extracted, as shown in **Figure 8a** (solid line) and approaches 2800 K. The light emission peak in the visible range rapidly increases with the applied electric field above a critical field ($\sim 0.4\ V / \mu m$). This is attributed to the accumulation of hot electrons and optical phonons in suspended graphene, above the critical field, which is related to the activation of intrinsic graphene optical phonons under an electric field. In suspended graphene, low activation to generate intrinsic optical phonons in an electric field are achieved by the reduction of energy loss suffered by electric field-induced hot electrons with extrinsic scattering sources, and prevent the direct cooling of hot electron and phonons through heat transfer to substrate. Current saturation in suspended graphene is observed and is known to be a signature of strong electron scattering by intrinsic optical phonons of graphene itself.

Raman spectroscopy provides an accurate measurement of acoustic phonons, optical phonon population and the temperature under applied electric field, as described in Sections 2.2 and 2.3. However, bright thermal radiation from electrically biased suspended graphene becomes significantly stronger than the Raman signal above ~1500 K. In order to estimate the temperature profiles and thermal conductivity, self-consistent numerical simulation of electrical and thermal transport is used, as described in Section 4. From numerical simulation based on the electrical transport data, thermal conductivity at the centre of the suspended graphene channel decreases from $k \sim 2700\ W / mK$ ($T_{ap} \sim 300$ K) to $k \sim 65\ W / mK$ ($T_{ap} \sim 1800$ K), as shown in **Figure 8c**. An abrupt decrease of thermal conductivity of suspended graphene also suppresses lateral heat dissipation along the graphene channel, meaning that hot electrons are confined and spatially localized at the centre of the graphene. As shown in **Figure 8d**, the maximum temperature is located at the centre of the graphene channel, which is consistent with the brightest light emission position as shown in **Figure 7**.

Effectively localization of hot electron in suspended graphene by unique electronic and thermal properties enables bright visible light emission. Based on the Stefan–Boltzmann law from measured electron temperature, as shown in **Figure 8e**, thermal radiation efficiency was estimated as ~ 4.45×10^{-3}, which is a 1000-fold enhancement compared to the graphene devices on substrates (~10^{-6}). Furthermore, engineering of radiation spectrum enhancement in the visible range provides the further enhancement of radiation efficiency.

An electrically driven graphene light emitter has great advantages, such as being atomically thin, broadband emission, radiation spectrum tunability, surface and self-emission all with a simple structure. Furthermore, ultrafast and broadband electrical and optical response of graphene has enabled the development of a nanoscale ultrafast light emitter. We expect to obtain atomically thin, flexible and transparent light sources and on-chip optical interconnects for communications based on graphene.

Acknowledgements

The authors thank J. Hone, Y.D. Park and G. Arefe. YDK was supported by Grants from ONR (N00014-13-1-0662 and N00014-13-1-0464) and DE-SC0012592. MB was supported by Grants from the National Research Foundation of Korea (NRF-2012-M3C1A1-048861, NRF-2015R1A2A1A10056103) funded by the Korean government.

Author details

Young Duck Kim[1*] and Myung-Ho Bae[2*]

*Address all correspondence to: yk2629@columbia.edu and mhbae@kriss.re.kr

1 Department of Mechanical Engineering, Columbia University, New York, NY, USA

2 Korea Research Institute of Standards and Science, Daejeon, Republic of Korea

References

[1] A K Geim and K S Novoselov. The rise of graphene. Nat Mater. 2007;6(3):183–191. doi: 10.1038/nmat1849

[2] Changgu Lee, Xiaoding Wei, Jeffrey W Kysar, and James Hone. Measurement of the elastic properties and intrinsic strength of monolayer graphene. Science. 2008;321(5887):385–388. doi:10.1126/science.1157996

[3] Alexander A Balandin, Suchismita Ghosh, Wenzhong Bao, Irene Calizo, Desalegne Teweldebrhan, Feng Miao, and Chun Ning Lau. Superior thermal conductivity of single-layer graphene. Nano Lett. 2008;8(3):902–907. doi:10.1021/nl0731872

[4] Y M Lin, C Dimitrakopoulos, K A Jenkins, D B Farmer, H Y Chiu, A Grill, and Ph Avouris. 100-GHz transistors from wafer-scale epitaxial graphene. Science. 2010;327(5966):662–662. doi:10.1126/science.1184289

[5] Changyao Chen, Sunwoo Lee, Vikram V Deshpande, Gwan-Hyoung Lee, Michael Lekas, Kenneth Shepard, and James Hone. Graphene mechanical oscillators with tunable frequency. Nat Nanotechnol. 2013;8(12):923–927. doi:10.1038/NNANO.2013.232

[6] Eric Pop. Energy dissipation and transport in nanoscale devices. Nano Res. 2010;3(3):147–169. doi:10.1007/s12274-010-1019-z

[7] Eric Pop, David Mann, Jien Cao, Qian Wang, Kenneth Goodson, and Hongjie Dai. Negative differential conductance and hot phonons in suspended nanotube molecular wires. Phys Rev Lett. 2005;95(15):155505. doi:10.1103/PhysRevLett.95.155505

[8] Marcus Freitag, Mathias Steiner, Yves Martin, Vasili Perebeinos, Zhihong Chen, James C Tsang, and Phaedon Avouris. Energy dissipation in graphene field-effect transistors. Nano Lett. 2009;9(5):1883–1888. doi:10.1021/nl803883h

[9] Stéphane Berciaud, Melinda Y Han, Kin Fai Mak, Louis E Brus, Philip Kim, and Tony F Heinz. Electron and optical phonon temperatures in electrically biased graphene. Phys Rev Lett. 2010;104(22):227401. doi:10.1103/PhysRevLett.104.227401

[10] Marcus Freitag, Hsin-Ying Chiu, Mathias Steiner, Vasili Perebeinos, and Phaedon Avouris. Thermal infrared emission from biased graphene. Nat Nanotechnol. 2010;5(7):497–501. doi:10.1038/nnano.2010.90

[11] Insun Jo, I-Kai Hsu, Yong J Lee, Mir Mohammad Sadeghi, Seyoung Kim, Stephen Cronin, Emanuel Tutuc, Sanjay K Banerjee, Zhen Yao, and Li Shi. Low-frequency acoustic phonon temperature distribution in electrically biased graphene. Nano Lett. 2011;11(1):85–90. doi:10.1021/nl102858c

[12] Myung-Ho Bae, Zhun-Yong Ong, David Estrada, and Eric Pop. Imaging, simulation, and electrostatic control of power dissipation in graphene devices. Nano Lett. 2010;10(12):4787–4793. doi:10.1021/nl1011596

[13] Dong-Hun Chae, Benjamin Krauss, Klaus von Klitzing, and Jurgen H Smet. Hot phonons in an electrically biased graphene constriction. Nano Lett. 2010;10(2):466–471. doi:10.1021/nl903167f

[14] Myung-Ho Bae, Sharnali Islam, Vincent E Dorgan, and Eric Pop. Scaling of high-field transport and localized heating in graphene transistors. ACS Nano. 2011;5(10):7936–7944. doi:10.1021/nn202239y

[15] Jae Hun Seol, Insun Jo, Arden L Moore, Lucas Lindsay, Zachary H Aitken, Michael T Pettes, Xuesong Li, Zhen Yao, Rui Huang, David Broido, Natalio Mingo, Rodney S Ruoff, and Li Shi. Two-dimensional phonon transport in supported graphene. Science. 2010;328(5975):213–216. doi:10.1126/science.1184014

[16] Nicola Bonini, Michele Lazzeri, Nicola Marzari, and Francesco Mauri. Phonon anharmonicities in graphite and graphene. Phys Rev Lett. 2007;99(17):176802. doi:10.1103/PhysRevLett.99.176802

[17] I Calizo, F Miao, W Bao, C N Lau, and A A Balandin. Variable temperature Raman microscopy as a nanometrology tool for graphene layers and graphene-based devices. Appl Phys Lett. 2007;91(7):07913. doi:10.1063/1.2771379

[18] Mathias Steiner, Marcus Freitag, Vasili Perebeinos, James C Tsang, Joshua P Small, Megumi Kinoshita, Dongning Yuan, Jie Liu, and Phaedon Avouris. Phonon populations and electrical power dissipation in carbon nanotube transistors. Nat Nanotechnol. 2009;4(5):320–324. doi:10.1038/nnano.2009.22

[19] Michael Engel, Mathias Steiner, Antonio Lombardo, Andrea C Ferrari, Hilbert v Löhneysen, Phaedon Avouris, and Ralph Krupke. Light–matter interaction in a microcavity-controlled graphene transistor. Nat Commun. 2012;3:906. doi:10.1038/ncomms1911

[20] Young Duck Kim, Myung-Ho Bae, Jung-Tak Seo, Yong Seung Kim, Hakseong Kim, Jae Hong Lee, Joung Real Ahn, Sang Wook Lee, Seung-Hyun Chun, and Yun Daniel Park. Focused-laser-enabled p–n junctions in graphene field-effect transistors. ACS Nano. 2013;7(7):5850–5857. doi:10.1021/nn402354j

[21] Inanc Meric, Melinda Y Han, Andrea F Young, Barbaros Ozyilmaz, Philip Kim, and Kenneth L Shepard. Current saturation in zero-bandgap, top-gated graphene field-effect transistors. Nat Nanotechnol. 2008;3(11):654–659. doi:10.1038/nnano.2008.268

[22] K I Bolotin, K J Sikes, Z Jiang, M Klima, G Fudenberg, J Hone, P Kim, and H L Stormer. Ultrahigh electron mobility in suspended graphene. Solid State Commun. 2008;146(9–10):351–355. doi:10.1016/j.ssc.2008.02.024

[23] Kirill I Bolotin, Fereshte Ghahari, Michael D Shulman, Horst L Stormer, and Philip Kim. Observation of the fractional quantum Hall effect in graphene. Nature. 2009;462(7270):196–199. doi:10.1038/nature08582

[24] Vincent E Dorgan, Ashkan Behnam, Hiram J Conley, Kirill I Bolotin, and Eric Pop. High-field electrical and thermal transport in suspended graphene. Nano Lett. 2013;13(10):4581–4586. doi:10.1021/nl400197w

[25] Young Duck Kim, Hakseong Kim, Yujin Cho, Ji Hoon Ryoo, Cheol-Hwan Park, Pilkwang Kim, Yong Seung Kim, Sunwoo Lee, Yilei Li, Seung-Nam Park, Yong Shim Yoo, Duhee Yoon, Vincent E Dorgan, Eric Pop, Tony F Heinz, James Hone, Seung-Hyun Chun, Hyeonsik Cheong, Sang Wook Lee, Myung-Ho Bae, and Yun Daniel Park. Bright

visible light emission from graphene. Nat Nanotechnol. 2015;10(8):676–681. doi: 10.1038/NNANO.2015.118

[26] Tian Fang, Aniruddha Konar, Huili Xing, and Debdeep Jena. High-field transport in two-dimensional graphene. Phys Rev B. 2011;84(12):125450. doi:10.1103/PhysRevB.84.125450

[27] Eduardo V Castro, H Ochoa, M I Katsnelson, R V Gorbachev, D C Elias, K S Novoselov, A K Geim, and F Guinea. Limits on charge carrier mobility in suspended graphene due to flexural phonons. Phys Rev Lett. 2010;105(26):266601. doi:10.1103/PhysRevLett.105.266601

[28] Denis L Nika, Artur S Askerov, and Alexander A Balandin. Anomalous size dependence of the thermal conductivity of graphene ribbons. Nano Lett. 2012;12(6):3238–3244. doi:10.1021/nl301230g

[29] J Moser, A Barreiro, and A Bachtold. Current-induced cleaning of graphene. Appl Phys Lett. 2007;91(16):1653513. doi:10.1063/1.2789673

Chapter 5

Electrospun Graphene Oxide-Based Nanofibres

Izzati Fatimah Wahab, Saiful Izwan Abd Razak,
Nor Syuhada Azmi, Farah Nuruljannah Dahli,
Abdul Halim Mohd Yusof and
Nadirul Hasraf Mat Nayan

Additional information is available at the end of the chapter

http://dx.doi.org/10.5772/64055

Abstract

Graphene and graphene oxide have been the forefront of research in the field of carbon nanostructures. In this chapter, the authors attempt to review and discuss works on electrospinning of graphene oxide-based nanofibre materials, mainly focusing on their combination with polymeric materials. Explanation and insights towards their method of preparation, properties and applications will be discussed. The conclusion section will give some closing remarks regarding the future direction of graphene-based nanofibres.

Keywords: graphene oxide, electrospinning, nanofibres, nanocomposites, nanomaterials, polymer matrices

1. Introduction

The discovery of graphene around this decade changed the focus direction of material science into its unique and magnificent characteristics. Researchers throughout the world are competing in finding ways to synthesis and explore the possible applications of graphene-based materials. Graphene oxide (GO) is an intermediate material that has better hydrophilicity and easier modification due to the presence of oxygen functionalities, compared to graphene. A broad range of applications have been reported over the years, including tissue engineering, sensor, power storage, coating and photonics. Electrospinning is one of the techniques that make use of GO materials in nanofibrous composites. This state-of-the art technique has been very popular in facile 1-D nano-scaled fibres production and gives materials their excellent properties in relation to high specific surface area.

2. Graphene oxide

The novel nanostructure of graphene has been showing promising characteristics not only in fundamental science but also in variety of applications. Its remarkable chemical and physical properties urged into many attempts of synthesizing graphene in mass production. However, it fails to give adequate productivity for large-scale use [1]. Graphene is not easy to be handled when it comes to incorporating it with other materials and to transform it into other form. Current development in this field recognizes graphene oxide (GO) as the best alternative to alleviate this problem, by having a more solvent friendly structure [2]. The solubility of GO in water and other solvents is likely in favour to the large commercialization of this novel material.

2.1. Structure

GO is basically a graphene sheet with oxygen functional groups, either epoxy, carboxyl or hydroxyl. It can be easily exfoliated from graphite through oxidation and sonication processes. Since graphene is a one-atomic layer of graphite, the structure of GO also consists in exfoliated layers of graphite oxide. Simplistically, the structural process is shown in **Figure 1**. The carbon compounds of **Figure 1** are organized in a honey-comb structure. One sheet of graphene contains six-carbon-atom hexagons, bonded by strong π and σ bonds.

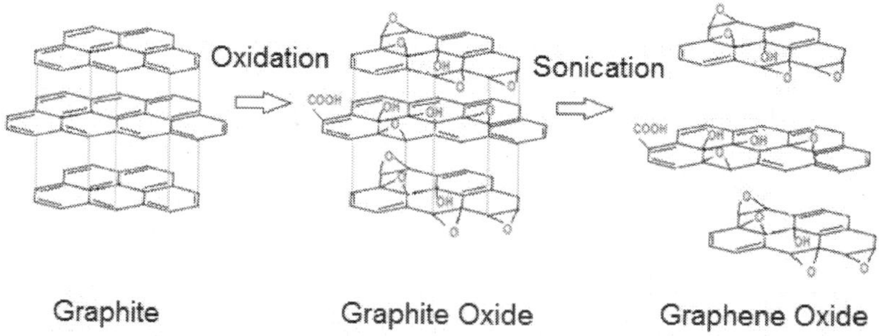

Figure 1. Structural process of graphene oxide.

The synthesis of GO involves oxidation of graphite into graphite oxide and exfoliation of graphite oxide into graphene oxide (**Figure 1**). There are several routes that can be utilized to oxidize graphite. The Brodie method, Brodie–Staudenmaier method and Hummers method are the earliest routes developed, but they lead to the release of toxic gases like NO_x and ClO_x [3]. Later, a typical modified Hummers method was introduced, and it has been used in many reported works in literature [4–6]. In general, graphite–graphene oxide conversion process can be summarized down into three steps [7]. First, the conversion of graphite into graphite intercalation compound (H_2SO_4–GIC), then, as a second step, H_2SO_4–GIC is converted into pristine graphite oxide (PGO). Finally, the conversion of PGO into GO takes place by the reaction of PGO with water.

The structure of GO on its own has become a research area interest among material scientists. This is due to the complexity of its non-stoichiometric composition [8]. The most recent and acceptable structure is constructed by Lerf and Klinowski. The Lerf–Klinowski model ignores the lattice-based model and focuses on a nonstoichiometric-amorphous alternative. It contains unoxidized benzene rings and aliphatic six-membered rings with epoxy, hydroxyl, carbonyl and/or carboxyl groups on the graphene sheet or at the graphene edges. Each oxygen atom is covalently bonded to two carbon atoms, forming an epoxy ring. Reduced graphene oxide (rGO) structure contains both sp^2 and sp^3 hybridized carbons and a high carbon/oxygen atoms ratio (C:O). sp^2 carbons bind to three other atoms, sharing a double bond with one of them, while a sp^3 carbon binds to a four other atoms.

A study on atomic and electronic structure of GO using annular dark field (ADF) imaging and electron energy loss spectroscopy (EELS) was conducted by Mkhoyan et al. in 2009 [9]. They found out that oxygen atoms randomly attach to graphene sites and convert sp^2 carbon bonds into sp^3 bonds. By ab initio calculated structure, two carbon atoms bonded to an oxygen atom are pulled above graphene plane. This configuration is fairly stable with −3.12 eV absorption energy. The bond length between these two carbon atoms expands to 1.514 Å from graphene 1.407 Å [9]. This is close to the diamond sp^3 bond length of 1.54 Å. The sp^3 hybridization and π and σ bonds of graphene oxide contribute the most to its unique physical and chemical properties. The extent of polar groups on the surface and edges of the graphene oxide depends on the amount of the oxidizing agent and the time and temperature of the oxidation reaction.

2.2. Properties

Spotted and closely studied only since 2004, graphene has already caught wide attention in research history after one by one of its interesting properties were reported. Being a single atomic layer of graphite, graphene is very thin and light. Yet, it is the strongest material ever discovered with intrinsic strength of 42 Nm^{-1} [10], so robust that it is 200 times stronger than steel and can outperform the hardness of diamond. It is transparent, but it can absorb light [11]. Having Young's modulus to approximately 1 TPa [10], it is elastic and pliable but, at the same time, is impermeable to all gases and liquids, except water [12]. It was also reported that graphene possesses superior thermal conductivity in the range of ~$(4.84 \pm 0.44) \times 10^3$ to $(5.30 \pm 0.48) \times 10^3$W/mK [13]. These values suggest that it can far exceed carbon nanotubes in conducting heat. Another important character of graphene is its exceptional electrical properties for having high intrinsic mobility of 2×10^5 $cm^2 v^{-1}s^{-1}$, compared to the highest known inorganic semiconductor (carbon nanotubes, with ~1×10^5 $cm^2 v^{-1}s^{-1}$) [14, 15].

Despite of being the building block of graphite layers, the synthesis of graphene is not a straight-forward exfoliation process. The method that discovered graphene sheets in 2004, using cellophane tape, does not appear to be a readily scalable process. Graphite exists in large quantities and is not expensive, yet producing sufficient graphene sheets is still a huge challenge [3]. Even the use of chemical vapour deposition (CVD) to deposit graphene on arbitrary substrates is not possible at low temperatures and is incompatible for mass production processes. Another concern is that graphene is a difficult material to be incorporated into various polymer matrices. Although graphene has unique properties, due to the strong π and σ bonds, these also contribute to its weak reactivity and low solubility in common organic solvents.

In order to prepare graphene–polymer composites or graphene-based hybrid materials, stable dispersion of graphene with other matrices must be obtained. This is when GO becomes useful. The sp^3 hybridization of GO contributes to a larger amount of free bonds, therefore increasing reactivity [16]. When graphite is oxidized, the functional groups make its hydrophilic and improve its exfoliation in aqueous media or organic solutions. The surface functionalization and solvents are the critical factors in determining interaction with functional groups of polymer or other matrixes. Several studies were done to determine the behaviour of GO towards a wide variety of solvents. Paredes et al. [17] studied the dispersibility of GO in acetone, methanol, ethanol, 1-propanol, ethylene glycol, dimethyl sulfoxide (DMSO), *N,N*-dimethylformamide (DMF), *N*-methyl-2-pyrrolidone (NMP), pyridine, tetrahydrofuran (THF), dichloromethane (DCM), o-xylene and n-hexane. They have successfully identified some type of solvents (DMF, NMP, THF and ethylene glycol) that can maintain a stable GO solubility. Later, another similar study compared GO and rGO solubility in eighteen different solvents. That investigation suggests that polarity of solvents is not the only factor for good dispersibility [2]. NMP and ethylene glycol, besides water, were double-proved to be the best solvents for long-term stability.

The covalent character of C–O bonds in GO disrupts the sp^2 conjugation of pure graphene and gives GO its insulating properties. Electrical conductivity can be recovered by restoring the π–network. GO, due to the presence of abundant oxygen, can be easily modified. Control of oxidation can provide tunable electronic properties, including the possibility of accessing zero-band gap graphene via bonds. GO, which contains a C/O content ratio of 2/1, is non-conductive (less than 1 S/m). GO can be partially reduced to conductive graphene-like sheets by removing the oxygen-containing groups. Restoration of the π–network and conjugated structure can be obtained by reducing GO to 6/1 of C/O atomic ratio. GO reduction is a process that converts sp^3 to sp^2 carbon, and thus recovers the important electrical properties. Some researchers make use of the oxygenated functionalities to bind with functional groups of conducting polymers, such as polyaniline, to improve the conductivity.

3. Electrospinning

At nanoscale, materials exist in 0, 1, 2, or 3 dimensions. One dimensional nanofibres can be synthesized by several techniques such as phase separation, self-assembly, template synthesis and electrospinning. Among those techniques, electrospinning has been recognized as a facile

http://dx.doi.org/10.5772/64055

process in fabricating polymeric fibres in nano-size [18]. It promptly gained popularity since its establishment, due to its potential implementation in various fields such as tissue engineering, chemical and biological sensing, energy generation and textiles.

3.1. Basics

The first observation on electrospinning dates back from the study of the electrospraying method used by Rayleigh (1897) and Zeleny (1914) [19]. Formhals, however, is the one who patented the electrostatic force experimental setup of polymer filaments production. In 1934, Formhals published a patent in which describing the production of textile yarn from cellulose acetate [20]. Since then, further research was carried out in order to assemble fibre processing using electrostatic force. Later, the study of electrically driven jets, by Taylor in 1969, became a strong foundation in understanding the electrospinning process. Only around 1994, the electrospinning term was used. Along with its popularity and rapid establishment in many fields, electrospinning has undergone many modifications and upgrading, in order to maximize its production and effectiveness in fabricating versatile nanofibres. There are several advanced techniques and setups of electrospinning instruments used favourably according to the required characterizations. Coaxial electrospinning, multilayer and mixed electrospinning, forced air-assisted electrospinning and air-gap electrospinning are a number of innovative processing techniques to enhance the function of electrospun fibres [21].

Basically, the typical basic components of any electrospinning setup are the high-voltage power supply, the spinneret (normally syringe with blunt-tip needle) and the collector (normally grounded). This basic experimental setup is shown in **Figure 2**. The polymer melt or solution placed into the syringe is held by retort stand (for vertical setups) or with a syringe pump (for vertical and horizontal setups). The tip of the syringe needle is connected to the positive electrode of the voltage supply. From here, the polymer liquid is pumped out of the syringe through the needle and brought to the grounded collector beneath the syringe or placed horizontally.

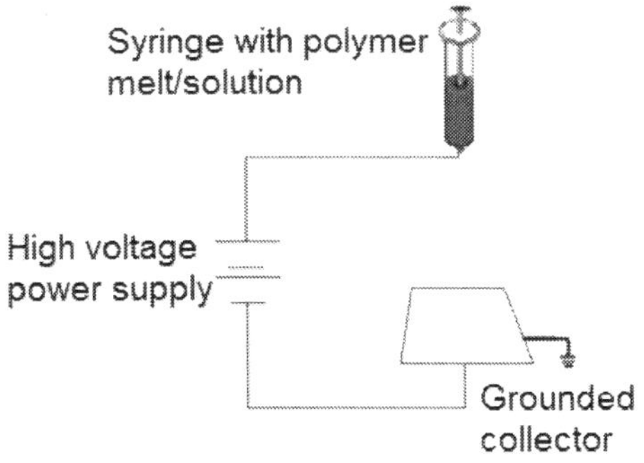

Figure 2. Electrospinning basic experimental setup.

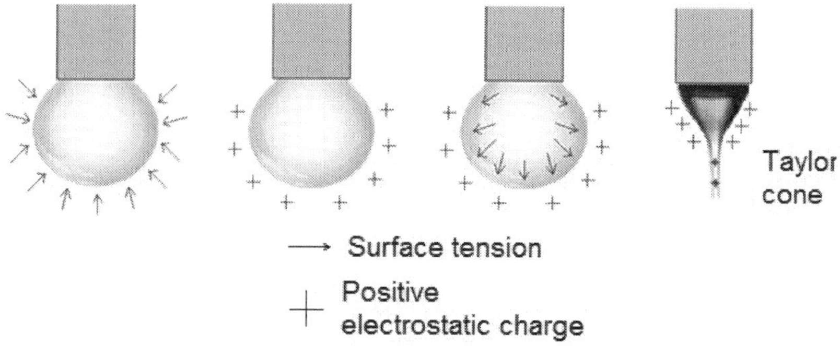

Figure 3. Description of electrospinning principle.

This unique approach for producing continuous fibres from polymer melt or solution, works by the principle of surface tension acting on the polymer drops. By the time, electricity is supplied onto the polymer, the fluid droplet will gain electrostatic charges, being positive charged. Due to the interaction between two forces, a Taylor cone is formed. The increase in high-voltage supply then eventually achieves a certain critical point, at which the electrostatic repulsion can overcome the surface tension holding the polymer droplet. It results in a liquid jet ejection targeting the collector. **Figure 3** shows a clearer depiction of the principle behind electrospinning technique.

The polymer jets onto grounded collector from the spinneret can be divided into two segments. There are streaming jet and whipping jet (**Figure 4**). The streaming jet can be seen with appropriate lighting, while the whipping jet is invisible to the naked eye. The flow of the streaming jet may give a wise initial prediction on the morphology of collected nanofibres. Leach et al., in 2011, indicated that streams behaving inconsistently, non-uniformly, in short

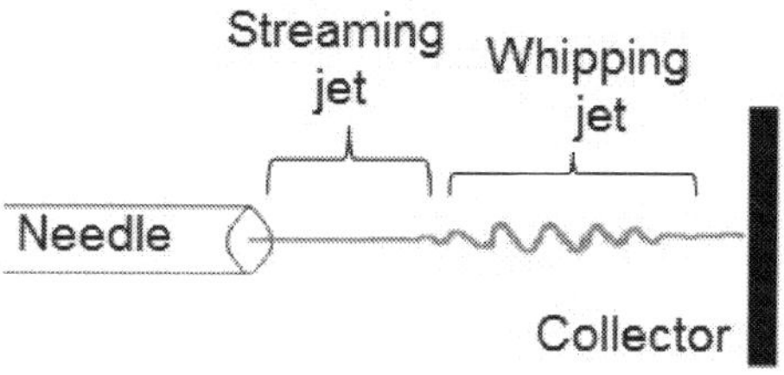

Figure 4. Streaming and whipping jets.

http://dx.doi.org/10.5772/64055

and oscillating, might produce fibres with poor alignment, beading, wavy pattern and splattering [22]. Therefore, researchers engaging in electrospinning studies need to carefully study various parameters (which will be further explained in the next section) before the optimized nanofibre formation procedure can be achieved.

3.2. Working parameters

Electrospinning deals with many working parameters in order to fabricate the required morphology and properties in the as-spun nanofibres. Other than the electrospinning experimental setup, like having different kind of collectors or spinnerets, electrospun nanofibres are also mainly affected by working parameters. The factors to be adjusted are the properties of polymer solution or melt, operational properties and ambient parameters [18, 22]. The electrospinning solution and operational properties are among the major parameters that influence electrospun nanofibres traits. Solution viscosity, surface tension and conductivity, along with electric field strength, distance between spinneret tip with collector and also the flow rate, are thoroughly investigated by researchers to maintain optimized fibre formation. Meanwhile, most researchers prefer to keep ambient parameters such as solution temperature, humidity and surrounding air velocity as constants throughout their studies. **Table 1** displays some factors taken into account for tuning the electrospun fibre morphology. Note that whenever finer or smaller diameter fibres are spun, they tend to get beading. Therefore, it is important for researchers to recognize their requirements and determine the optimized conditions for electrospun fibre formation. There are also some parameters that need minimum requirements to be executed in order to induce electrospinnability. The applied voltage and tip-to-collector distance, for example, are some of those parameters, having certain minimum

Parameter	Effect on fibre morphology
Viscosity	Fibre diameters increase with solution viscosity. Also, higher viscosity results in more beading
Concentration	High solution concentration results in larger fibre diameter
Applied voltage	Higher-voltage supply jets more fluid and cause larger fibre diameter Increase in electrical potential also results in rougher fibres
Surface tension	Reduction in surface tension minimizes fibre beads
Feed-rate	Low feeding rate: decrease in fibre diameter High feeding rate: increase in bead generation and Taylor cone instable
Tip-to-collector distance	Minimum distance required for dry fibres. Excessively far distance will lessen the deposited amount of fibres on the collector
Conductivity	Decrease in fibre diameter with increase in conductivity, bending instability and diameter distribution

Table 1. Effect of parameters on fibre morphology [19, 23].

values which differ for every electrospinning material. Other parameters that rule electrospinning process are the molecular weight of the material used, types of collector and of solvent used, and ambient parameters [18, 23].

4. Electrospinning of GOs with polymers

The establishment of GO incorporation into polymeric electrospun nanofibres was only initiated around this decade. Since 2010, at least 34 literature works were published involving about 14 types of synthetic and natural polymers. There are some distinct reasons to add GO and its derivatives into electrospun nanofibres. The major purpose is to improve the mechanical and electrical properties of the electrospun material. Others include the morphological enhancement to meet certain diameters or porosity of the spun fibres. **Table 2** summarizes the reported polymer/GO combinations.

Graphene oxide	Polymer	References
Graphene oxide	Poly (vinyl acetate) (PVAc)	[56]
Graphene oxide	Polycaprolactone (PCL)	[42, 56]
Functionalized graphene oxide	Poly (lactic acid) (PLA)	[26, 41]
Graphene oxide	poly(lactic-co-glycolic acid) (PLGA) and collagen (Col)	[40]
Graphene oxide	poly(lactic-co-glycolic acid) (PLGA)	[32, 40]
Graphene oxide	PLA and polyurethane (PU)	[38]
Graphene oxide	Polyurethane (PU)	[38, 53, 55]
Graphene oxide	Polyethylene oxide (PEO)/chitosan (CS)	[44]
Graphene oxide	Poly (vinyl alcohol) (PVA)	[31, 35, 44, 48]
Reduced graphene oxide	Poly (vinyl alcohol) (PVA)	[24, 27]
Graphene oxide	PVA and chitosan CS	[33, 47]
Graphene oxide	Nylon 6-6	[52]
Reduced graphene oxide	Nylon-6	[57]
Graphene oxide	Nylon-6	[34]
Reduced graphene oxide	Poly(carbonate urethane) (PCU)	[30]
Reduced graphene oxide	polyvinyl pyrrolidone (PVP)	[25]
Reduced graphene oxide	polyacrylonitrile (PAN)	[25, 29, 36]
Functionalized graphene oxide	polyacrylonitrile (PAN)	[28]
Graphene oxide	Polyacrylonitrile (PAN)	[28, 50]
Reduced graphene oxide	Polyvinyl Butyral (PVB)	[46]
Graphene oxide	Polyacrylic acid (PAA)	[54]

Table 2. Reported literatures on electrospun GO.

A fundamental study by Park et al. in 2013 described the effect of GO distribution and its location in 1D polymer composites. By controlling the distribution and location of the embedded nanomaterials in 1D nanoscale composites, the macroscopic properties can be enhanced and thus enable many practical applications. In their study, three types of polymers and two different solvents were used. The type of polymers and solvents are two processing conditions to manipulate the distribution and location of the GOs. Homogeneous colloidal dispersions of GO in the polymer solutions were successfully electrospun into polymer nanofibres, without the formation of beads or aggregates [24]. The location of GO was influenced by phase separation during the electrospinning process, which was in turn dependent on the compatibility between the two components and the vapour pressure of the solvents. The thermal and mechanical properties of polymeric composite nanofibres were effectively improved by adding GO as nanofiller.

GOs have a curved morphological structure and form powder materials after exfoliation from graphite oxides. This powder structure can be used in paints, adhesives, coatings and lubricants, where GOs act as multifunctional additives. However, for applications where structured thickness is a limitation, the levelling of the graphene sheets in a plane is required. Thus, an effort to study the levelling of graphene sheets through electrospinning was recently carried out [25]. Those authors used PAN and PVP as examples of polar polymers for nanofibre matrices. GO was reduced using different agents, but later it was confirmed that the curved features of GO sheets cannot be prevented using arGO coating strategy. Nonetheless, the strategy of using electrospinning to extend single GO sheets seems effective. Levelled graphene sheets were successfully developed and showed good conductivity in the order of 10^3 Sm^{-1}.

The effect of GO loading into PCL was investigated in presence of different oxidation levels [26]. GO dispersibility, wettability, polarity and charge relaxation time, which assess the fibre diameter, increase with oxidation level. Surface tension of PCL solution is invariable in the presence of GO, resulted in diameter variations of the spun fibres. The diameter of PCL nanofibres decreases with GO loading, due to the reduction of solution viscosity and increase in the conductivity. Another morphological study attempted to blend rGO into PVA [27]. The addition of a nanofiller seems to increase the viscosity and conductivity, but the nanofibre exhibited smaller diameters than those without nanofillers. Some GO particles were also observed to be protruded from the composite nanofibres under transmission electron microscope (TEM).

The manufacturing of commercial high-performance carbon nanofibre (CNF) has been carried out using PAN as precursor, due to its high carbon yield and superior fibre strength. It includes a typical process of oxidative stabilization and carbonization steps. Nevertheless, most continuous CNFs produced to date have relatively poor graphitic structure, which is important for higher elastic moduli and thermal and electrical conductivities. This can be solved by increasing the carbonization temperature, because high temperature graphitization significantly improves the size and alignment of the graphitic crystallites in PAN-based CNFs. This high-temperature condition is, however, unlikely since fibre processing at such conditions requires a specialized expensive equipment. Therefore, Papkov and co-workers, in 2013, carefully made an effort to maximize the orientation of the polymer chains inside the polymer precursor fibre by incorporating oriented inclusions that had high surface area and could strongly interact with polymer chains [28]. GO was used since it can both organize the polymer chains and simultaneously serve as a 'templating' agent for the formation of graphitic crystallites during carbonization. Their close examination revealed the presence of crumpled GO particles inside the PAN matrix. The irregular shape of the radially crumpled sheets closely matches the popular model of the high-strength carbon fibre structure, which consists of irregular, longitudinally wrinkled turbostratic graphitic stacks with preferred axial orientation. It is very useful to represent a natural template to achieve similar high-strength CNFs structure in economic manufacturing protocol. In addition, the crumpled graphene oxide particles may provide multiple nucleation sites for graphitic crystal growth during carbonization, especially because the carbon atoms along the crumpled folds are more reactive.

The paramount role of graphene-based embedded in polymer matrix via electrospinning is to enhance the mechanical properties of the polymeric nanofibres. Wang et al., in 2015, examined PAN/GO composite nanofibres with lateral force microscopy and force-distance curves, in order to assess the surface properties like friction force and elasticity. The experiment results indicated that, with increasing GO concentrations, both the surface friction force and adhesive force increased [29]. A mechanical characterization of high-performance GO incorporated into an aligned fibro-porous PCU membrane was made, under static and dynamic conditions [30]. The electrospun membrane indicated that 55% of the tensile strength increased, a 127% rise in toughness, and the achievement of maximum strength reinforcement efficiency was reported at 1.5 wt.% GO loading. An even smaller loading of GO (0.02%) in PVA diluted aqueous solution was able to produce elevation of tensile strength of electrospun nanofibres by 42 times [31].

Another study on significant improvements in thermomechanical and surface chemical properties of poly(D, L-lactic-co-glycolic acid) (PLGA) upon addition of GO nanosheets was published in 2011 [32]. A significant enhancement of storage moduli and glass transition temperature of the PLGA/GO (2 wt.%) was presumably reported to be caused by the enhanced chemical bonding between the GO oxygenated functional groups and the polymer hydroxyl groups. This nanocomposite fibre was more hydrophilic than the pristine PLGA fibre. Biocompatibility tests showed that the enhanced surface chemical properties of the PLGA/GO (2 wt.%) effectively improved neuronal cell proliferation and viability, indicating enhanced biocompatibility of PLGA nanocomposite nanofibres by the addition of 2-D GO nanofillers. Novel electrospun PVA embedded with glycine-modifed GO (GMGO/PVA) and PVA with nanogold deposited GO (AuG/PVA) was fabricated and its biocompatibility was critically analysed [33]. GMGO was synthesized by reducing GO with a biocompatible amino acid, glycine, for conductivity improvement. Then, AuG was modified from GMGO, by mixing with hydrogen tetrachloroaurate ($HAuCl_4$). Cellular toxicity tests of these nanofibres showed almost no cytotoxicity to L929 cells, most likely due to nontoxicity of PVA nanofibres and the biocompatibility of modified GO.

http://dx.doi.org/10.5772/64055

There were also some experiments carried out with produce electrospun GO/polymer nanofibres for morphological modification, in order to meet certain requirements for applications. A spider-wave-like nano-net was fabricated from GO blend with nylon-6 which holds polyelectrolytic and polymorphic characteristics [34]. The amide (CO-NH) group of nylon-6 molecule can easily interact with GO by means of hydrogen bond formation or donor-acceptor complexes. The strategy was to incorporate foreign materials into nylon-6 in order to change the morphology and crystallization of nanofibres. The pristine nylon-6 electrospun mat appears well defined with no spider-wave-like nano-net. As GO was added into the polymeric solution, a double-layered morphology (bimodal fibre diameter distributed mat) was noticeable. Further increase of GO in the polymeric solution not only decreased the spider-wave-like structure but also decreased the spinnability and formed a film-like structure.

The widespread use of electrospinning technique in tissue engineering application is no longer new. Many researchers have been exploring this subject. GO/polymer composites also found their way through this area, since they showed potential biocompatibility and noncytotoxicity characteristics. Scaffolds play an important role in tissue engineering as they not only provide support for cell attachment, but they also allow cell growth into tissue until the tissue is able to sustain itself. There are several types of GO/polymer composites been developed for tissue engineering scaffolds, such as PVA/GO [35], PAN/rGO [36], PVA/CS/GO [37], PLA/PU/GO [38], PLA/hydroxyapatite (HA)/GO [39], GO/PLGA/Col [40], GO-g-PEG/PLA [41] and PCL/GO [42]. The polymeric scaffolds attain better thermal, mechanical and electrical properties, due to the intrinsic properties of GO. Cells adhesion and proliferation onto these nanofibrous scaffolds have been tested morphologically and quantitatively. Cells like mouse osteoblastic cells (MC3T3-E1), adipose-derived stem cells (ADSCs), skeletal myoblasts (C2C12), mouse marrow mesenchymal stem cells (MSCs) and low-differentiated rat pheochromocytoma cells (PC12-L) were cultured onto the samples. They confirmed that a small amount of GO did not restrain the proliferation and viability of cells, which demonstrated the appreciable cell affinity of GO. Cells are likely to spread on the composite scaffolds and some even enhance the cell growth by having similar rate of cell proliferation to that of tissue culture plates.

Tissue engineering is very important for regeneration and replacement of damaged tissues in the human body. Scaffolds for musculoskeletal and skin tissues have caught the attention of researchers in the last years, aiming to help many patients with injuries. On the other hand, tissue engineering for organs or tissues without regenerative abilities is an extensive area that needs serious exploration. Vascular tissue engineering is one of its branches, and it aims to fabricate functional vascular grafts to be used in vivo, in order to replace blood vessels and help their regeneration. Current commercial synthetic grafts used by surgeons are woven poly (ethylene terephthalate) (Dacron) and extended polytetrafluoroethylene (ePTFE). Although the aim was to substitute large diameter blood vessels, they are not suitable for small diameter grafts due to thrombosis and intimal hyperplasia risks. A thermoplastic polyurethane (TPU)/GO scaffold was fabricated using grounded rotation mandrel as electrospinning collector to obtain fibrous small diameter tubular scaffolds [43]. The tensile strength

(Young's modulus) and scaffolds hydrophilicity increased as GO amount increased. The suture retention strength and burst pressure of tubular TPU/GO scaffolds containing 0.5 wt. %. GO exceeded the mechanical requirements of human blood vessels. In addition, results of platelet adhesion tests indicated that a low loading of GO is unlikely to cause thrombosis on the scaffolds. Endothelial cells could form a layer on the inner surface of the tubular scaffolds which resembles the native blood vessel structure, indicating that the TPU/GO electrospun scaffolds, with low GO loading level, have the potential to be used in small diameter vascular graft applications.

Another impressive function of scaffolds that needs to be expanded is the ability to **control drug release** onto specific tissues or organs. A targeted drug delivery system that can efficiently transport drugs is a way for a scaffold to improve the efficacy of the drug and to reduce the systemic toxicity. Electrospun nanofibres are very important materials due to the high surface area and high porosity of the nanofibres, which are able to provide higher drug encapsulation efficiency and better stability, compared to other systems. An electrospun scaffold for anticancer drug delivery was tested using a polyethylene oxide (PEO)/chitosan (CS)/GO nanocomposite [44]. In the nanofibre carrier system, GOs act as nanocarriers which control doxorubicin (DOX) release, an anthracycline antibiotic used in cancer chemotherapy intravenous administration. DOX and GO surface can make a strong bond through π–π stacking interactions as a non-covalent type of functionalization that may provide controlled release of drug. GO also contains epoxy and carboxylic functional groups that can react with the amine groups of CS. The π–π stacking interaction between DOX and GO with the fine pores of nanofibrous scaffolds allowed high drug loadings (98%).A strong pH dependence was shown by the DOX release result at pH 5.3 and 7.4. At pH 5.3, faster drug release can be seen which may resulted from the unstability of hydrogen-bonding interaction between GO and DOX.

Along with the development of tissue engineering field, the **fabrication of scaffolds** is also advancing and succeeding in providing a mimic environment to the real body system. Currently, combining biomaterial with living cells for tissue scaffolds is a critical approach. Luo et al., in 2015, have reported a fabrication of GO-doped ply (lactic-co-glycolic acid) (PLGA) nanofibrous scaffold mixed with human MSCs for osteogenic differentiation [45]. The porous structure and the diameter of produced fibres are similar to the topological structure of natural extracellular matrices (ECM), and therefore facilitating cell attachment and proliferation. GO plays two roles in this scaffold. First, it enhances the hydrophilic performance and protein- and inducer-adsorption ability of the nanofibres. Second, it accelerates the human MSCs adhesion, proliferation and differentiation towards osteoblasts.

Graphene-based materials have extraordinary electronic transport properties, strong mechanical strength, excellent electrocatalytic activities, high surface area, ease of functionalization, and thus have contributed to ion detector and **biosensing** applications. A sensor for Cu (II) detector based on rGO/polyvinyl butyral (PVB) nanofibres was fabricated by Ding and co-workers in 2015. Electrospun GO/PVB on glassy carbon electrode (GCE) was reduced through an electrochemical method, to form reduced graphene oxide (RGO)/PVB nanofibres [46]. Cu (II) is one of the heavy metals that contaminate water sources through casual discharge from mining, machinery manufacturing and metal smelting. Although copper is an essential element in human body, the excessive intake of copper might results in Wilson's disease and Menke's syndrome [46]. The rGO/PVB nanofibres modified glassy carbon electrode (GCE) were used for Cu (II) detection by differential pulse anodic stripping voltammetry (DPASV) [46]. The as-fabricated sensor based on rGO/PVB nanofibres showed good analytical performance with the linear range of 0.06–22 μM, a low detection limit of 4.1 nM (S/N=3), high sensitivity of 103.51 μA •μM^{-1} cm^{-2}, good selectivity and excellent reproducibility (RSD = 0.49%) [46].

A novel platform for enhanced biosensing of electrospun PVA/CS/GO was also engineered [47]. Authors used glucose oxidase (GOD) as a model enzyme. Nanofibres produced from electrospinning are an ideal biosensing platform because of the large surface area they provide, leading to high enzyme loading capacity and low hindrance for mass transfer. Electrospinning is also a simple, durable and adjustable prove to provide stability and reproducibility of biosensors. GOs can bind enzyme molecules well, through electrostatic interactions, and thus avoiding the leaching of enzymes from modified electrodes. Nanomaterials are needed in nanofibre platform to bind the enzyme, since accessibility of the substrate to enzyme will be inhibited when the enzyme is confined inside the nanofibres. The fabricated biosensing platform displayed optimal oxidation and reduction capabilities towards electro-active species, even when compared to the bare Pt electrode, due to the synergy effects of PVA/chitosan nanofibres and GO nanosheets [47].

Power generation is one of the essential needs of human daily life. The increasing demand for green and rechargeable energy sources to power up many electronic devices urged into a variety of power storage system alternatives. Knowing the superior properties of graphene and high specific surface area of CNFs, a number of attempts have been made to incorporate these materials into the fabrication of electrodes for Li-batteries and supercapacitors. Current freestanding anode for Li-ion batteries of graphite has limited theoretical capacity (372 mA h g^{-1}). Xu et al., in 2014, proposed the use of Si, a nontoxic and abundant element, as the anode, due to its highest ever-known theoretical capacity up to 4200 mA h g^{-1} [48]. However, Si has its limitations, as it can undergo up to 280% volumetric expansion upon lithiation, which leads to electrode pulverization and loss of electrical contacts, resulting in severe capacity fading. The solution to this problem is to convert bulk Si into particles of low dimensions and to uniformly disperse Si in an inactive and conductive matrix, like GO. Co-axial nozzle was used to electrospun Si/CNF/GO anode, in order to avoid agglomeration during the spinning process. The resulting electrode achieved a high rate capacity of 567 mA h g^{-1} at 1 mA h g^{-1} current density. It obtained excellent cyclic performance with a discharge capacity of 872 mA h g^{-1}after 50 cycles and 91% capacity retention. Another approach is the synthesis of core-shell structured nanofibres that consist of rGO/PAN (shell) and ZnO/carrier polymer (core) [49]. The current use of ZnO has actually gained a good theoretical capacity of 978 mA h g^{-1}. Unfortunately, it also carries a huge volume expansion for approximately 228%, which causes damage of mechanical integrity. Considering the same solutions, Shilpa et al., in 2015, encapsulated ZnO nanoparticles in the hollow core of glassy carbon–reduced graphene oxide (C–rGO) where the void spaces should provide a buffer zone, accommodating the volume changes associated with lithiation/delithiation of the metal oxide and prevent attrition of nanoparticles by detachment and fragmentation. The outcome of this study gained capacitance of 815 mA h g^{-1} at a current density of 50 mA g^{-1} with a capacity retention of almost 80% after 100 cycles [49]. A freestanding electrode eliminates the use of current collector, which is an inactive material that only adds to the battery mass and volume.

Another CNF composite in form of fibre mats was prepared by electrospinning blend of GO and PAN in *N,N*-dimethylformamide (DMF) for supercapacitor electrodes [50]. Porous carbon material of CNFs was further improved by graphene-based electrical conductivity, through carbonization, to obtain CNF/graphene composites. GO concentration was the major factor responsible for the morphology and pore structures, where the optimized-specific surface area and conductivity were achieved at 5 wt.%. The electrode material (5 wt.% GO) exhibited enhanced electrochemical performance with a highest specific capacitance of 146.62 F g^{-1}, and highest energy density of 14.83–19.44 W h kg^{-1} in the power density range of 400–20,000 W kg^{-1}. Over the last decades, there was another upcoming power storage system that has been successfully applied in numerous fields. **Latent heat storage** using phase change material (PCM) makes use of the solid–liquid phase change of a material in order to store thermal energy. Fatty acid esters showed potential significant energy storage ability, as organic PCMs, with satisfactory thermal properties, thermal conductivity and thermal reliability. Nevertheless, the use of organic PCMs may undergo energy leakage. The synthesis of PCM with a supporting material is crucial to overcome the leakage problem and improve its thermal conductivity. Ke et al., in 2014, report on the preparation of methyl stearate (MES) fatty acid ester with PAN/GO by electrospinning [51]. GO was selected as the heat transfer/diffusion promoter and also as supporting material in the phase change composite. The results suggest that by adding GO, beneficial effects on mechanical properties, thermal stability and thermal energy storage/release performance were gained.

A work has been published dealing with the use of nylon 6-6 excellent mechanical resistance, with functionalized GO as a **corrosion protection** composite coating [52]. 2% GO loading in the electrospun nylon 6-6 possessed coating capacitance in the order of 10^{-7} F/cm^2, compared to 10^{-4} F/cm^2 for coating prepared by deposit. This value suggests charge storage capacity for the coating condition. Thus, the efficiency of deposited coating is better than electrospun coating. The well-known physicochemical properties of PU have been widely used in coating technology. However, its low hydrophilicity and mechanical properties are the main setbacks for its application. Meanwhile, GO-polymer composite fibres emerged as the subject of enormous interest because polymer nanocomposites, with graphene materials as fillers, have shown dramatic improvements in such needed properties. Electropinning of in situ synthesized PU/GO has been conducted for surface coating of metallic stents [53]. This study showed that the incorporation of small amounts of GO sheets through PU nanofibres could enhance the stability of coated layer on the surface of nonvascular stent. The grafted PU chains on the surface of GO can facilitate the beneficial stress transfer from the PU matrix to the GO sheets. Moreover, proper incorporation of GO sheets into the polymer matrix can improve the hydrophilicity of the materials, without any adverse biological effect. These improved properties of PU mat, induced by GO, make it a potential candidate for surface coating of medical devices.

In order to fabricate **water-swellable rubber** (WSR), a multi-scale hybrid nanofibre with enhanced elasticity, resilience, toughness and water swelling ability was electrospun [54]. WSR can be produced by chemical and physical methods. The chemical method is more complicated and more environmentally unfriendly than the physical method. In the physical method, common hydrophobic rubbers are mixed with various water-absorbent materials, known as superabsorbent polymers (SAPs) (hydrogel form). Hydrophilic super water-absorbent resin does not disperse well in hydrophobic rubber, expecting easy break off of physical-synthesized WSR. Eventually, swelling ability will lose, strength is remarkably decreased, and water is polluted. The addition of hydrophilic fibres can fix then as they can act as internal water channels in hydrophobic materials for superabsorbent nanocomposites. These fibres can help to transfer water from the surface of the rubber matrix to the hydrophilic SAP particles, as well as between encapsulated SAP particles in the rubber matrix, enhancing the water swelling ability of WSR composites. Fabrication of PAA/ethylene glycol (EG)/hyperbranched bis-MPA polyester-64-hydoxyl (HB)/GO showed synergistic effects on water swelling ability compared to without HB and GO. The electrospun fibre mats were added into conventional rubber by simple physical mixing, and the rubber composites showed considerable enhancement of water swelling ability, mainly due to the water channels built.

Functional **shape memory** composite nanofibres with GO have also been fabricated with PU shape memory [55]. Shape memory PU/GO (SMPU/GO) has better shape memory effect (SE) and lower thermal shrinkage, compared to SMPU nanofibre mats. When the loading amount of GO increased to 4 wt.%, the thermal shrinkage ratio of the composite nanofibrous mats reached as low as 4.7 ± 0.3%, while the average fixation ratio and recovery ratio were as high as 92.1 and 96.5%, respectively. The study indicates that GO is a desirable reinforcing filler for preparing shape memory nanofibres with improved properties. Electrospun polymer with GO and modified GO fillers have been extensively investigated for a wide range of applications. Apart from all applications mentioned, electrospinning of polymer/GO also stands out as an optical element in fibre lasers [56] and photocatalysts [57], showing their important role in the area of photonics.

5. Conclusions

Graphene oxide and its derivatives show comparable properties as alternatives to graphene. Their outstanding performance can easily be manipulated into a wide variety of applications. An important highlight is that several researchers have shown that the biocompatibility of produced composites is still preserved, with no significant cytotoxicity appearing at low GO loading. Utilizing electrospinning to fabricate polymer/GO composites is a wise choice to obtain high specific surface area fibres at the nanoscale. It displays a promising advantage in achieving optimum properties out of the electrospun composites. Nonetheless, further research towards commercializing polymer/GO electrospun fibres in applications and improvement on electrospinning machines needs to be done, so that one can make use of their outstanding advantages to a larger extent.

Author details

Izzati Fatimah Wahab[1,2], Saiful Izwan Abd Razak[1,2*], Nor Syuhada Azmi[2], Farah Nuruljannah Dahli[3], Abdul Halim Mohd Yusof[3] and Nadirul Hasraf Mat Nayan[4]

*Address all correspondence to: saifulizwan@utm.my

1 IJN-UTM Cardiovascular Engineering Centre, Universiti Teknologi Malaysia, Skudai, Johor, Malaysia

2 Faculty of Biosciences and Medical Engineering, Universiti Teknologi Malaysia, Skudai, Johor, Malaysia

3 Faculty of Chemical and Energy Engineering, Universiti Teknologi Malaysia, Skudai, Johor, Malaysia

4 Faculty of Engineering Technology, Universiti Tun Hussein Onn Malaysia, Batu Pahat, Johor, Malaysia

References

[1] D. Li, M. B. Mueller, S. Gilje, R. B. Kaner, and G. G. Wallace, Processable aqueous dispersions of graphene nanosheets, Nat. Nanotechnol., vol. 3, no. 2, pp. 101–105, 2008.

[2] D. Konios, M. M. Stylianakis, E. Stratakis, and E. Kymakis, Dispersion behaviour of graphene oxide and reduced graphene oxide, *J. Colloid Interface Sci.*, vol. 430, pp. 108–112, 2014.

[3] J. Li, X. Zeng, T. Ren, and E. van der Heide, The preparation of graphene oxide and its derivatives and their application in bio-tribological systems, *Lubricants*, vol. 2, pp. 137–161, 2014.

[4] G. L. Li, G. Liu, M. Li, D. Wan, K. G. Neoh, and E. T. Kang, Organo- and water-dispersible graphene oxide-polymer nanosheets for organic electronic memory and gold nanocomposites, *J. Phys. Chem. C*, vol. 114, no. 29, pp. 12742–12748, 2010.

[5] T. Kuilla, S. Bhadra, D. H. Yao, N. H. Kim, S. Bose, and J. H. Lee, Recent advances in graphene based polymer composites, *Prog. Polym. Sci.*, vol. 35, no. 11, pp. 1350–1375, 2010.

[6] A. Fazal and K. S. Fancey, Performance enhancement of nylon kevlar fiber composites through viscoelastically generated pre-stress, *Polym. Polym. Compos.*, vol. 16, no. 2, pp. 101–113, 2008.

[7] A. M. Dimiev, J. M. Tour, Mechanism of graphene oxide formation, *ACS Nano*, vol. 8, no. 3, pp. 3060–3068, 2014.

[8] D. R. Dreyer, S. Park, C. W. Bielawski, and R. S. Ruoff, The chemistry of graphene oxide, *Chem. Soc. Rev.*, vol. 39, no. 1, pp. 228–240, 2010.

[9] K. A. Mkhoyan, A. W. Contryman, J. Silcox, D. A. Stewart, G. Eda, C. Mattevi, and S. Miller, Atomic and electronic structure of graphene-oxide, *NANO Lett.*, vol. 9, no. 3, pp. 1058–1063, 2009.

[10] C. Lee, X. Wei, J. W. Kysar, and J. Hone, Measurement of the elastic properties and intrinsic strength of monolayer graphene, *Science (80-.).*, vol. 321, no. July, pp. 385–388, 2008.

[11] R. R. Nair, A. N. Grigorenko, P. Blake, K. S. Novoselov, T. J. Booth, N. M. R. Peres, T. Stauber, and A. K. Geim, Fine structure constant defines visual transparency of graphene., *Science (80-.).*, vol. 320, no. 5881, p. 1308, 2008.

[12] J. S. Bunch, S. S. Verbridge, J. S. Alden, A. M. Van Der Zande, J. M. Parpia, H. G. Craighead, and P. L. Mceuen, Impermeable atomic membranes from graphene sheets, *NANO Lett.*, vol. 8, no. 8, pp. 2458–2462, 2008.

[13] A. A. Balandin, S. Ghosh, W. Bao, I. Calizo, D. Teweldebrhan, F. Miao, and C. N. Lau, Superior thermal conductivity of single-layer graphene, *Nano Lett.*, vol. 8, pp. 902–907, 2008.

[14] J. H. Chen, C. Jang, S. Xiao, M. Ishigami, M. S. Fuhrer, Intrinsic and extrinsic performance limits of graphene devices on SiO_2, *Nature Nanotechnol.*, vol. 3, pp. 206–209, 2008.

[15] T. Dürkop, S. A. Getty, E. Cobas, and M. S. Fuhrer, Extraordinary mobility in semiconducting carbon nanotubes, Nano Lett., vol. 4, pp. 35–39, 2004.

[16] G. Peschel, Carbon-carbon bonds: hybridization. Obtained online from: http://www.physik. fu-berlin.de/einrichtungen/ag/ag-reich/lehre/Archiv/ss2011/docs/Gina_Peschel-Handout. pdf, published on, 5(5), 2011.

[17] J. I. Paredes, S. Villar-Rodil, A. Martínez-Alonso, J. M. D. Tascón, Graphene oxide dispersions in organic solvents, *Langmuir*, vol. 24, pp. 10560–10564, 2008.

[18] C.-L. Zhang and S.-H. Yu, Nanoparticles meet electrospinning: recent advances and future prospects, *Chem. Soc. Rev.*, vol. 43, no. 13, pp. 4423–48, 2014.

[19] N. Bhardwaj and S. C. Kundu, Electrospinning: a fascinating fiber fabrication technique, *Biotechnol. Adv.*, vol. 28, no. 3, pp. 325–347, 2010.

[20] A. Formhals, Process and apparatus for preparing artificial threads, 1934. US Patent, 1975504.

[21] K. Garg and G. L. Bowlin, Electrospinning jets and nanofibrous structures, *Biomicrofluidics*, vol. 5, no. 1, pp. 1–19, 2011.

[22] M. K. Leach, Z.-Q. Feng, S. J. Tuck, and J. M. Corey, Electrospinning fundamentals: optimizing solution and apparatus parameters, *J. Vis. Exp.*, no. 47, p. e2494, 2011.

[23] Z. M. Huang, Y. Z. Zhang, M. Kotaki, and S. Ramakrishna, A review on polymer nanofibers by electrospinning and their applications in nanocomposites, *Compos. Sci. Technol.*, vol. 63, no. 15, pp. 2223–2253, 2003.

[24] H. S. Park, B. G. Choi, W. H. Hong, and S.-Y. Jang, Controlled assembly of graphene oxide nanosheets within one-dimensional polymer nanostructure, *J. Colloid Interface Sci.*, vol. 406, pp. 24–29, 2013.

[25] Y. Wang, J. Tang, S. Xie, J. Liu, Z. Xin, X. Liu, and L. A. Belfiore, Leveling graphene sheet through electrospinning and its conductivity, *RSC Adv.*, vol. 5, no. 52, pp. 42174–42177, 2015.

[26] S. Ramazani and M. Karimi, Electrospinning of Poly(e-caprolactone) solutions containing graphene oxide: effects of graphene oxide content and oxidation level, *Polym. Compos.*, vol. 37, no. 1, pp. 131–140, 2014.

[27] C. L. Huang, S. Y. Peng, Y. J. Wang, W. C. Chen, and J. H. Lin, Morphology of Poly(vinyl alcohol)/Graphene Nanosheets Composite Nanofibers, *Appl. Mech. Mater.*, vol. 365–366, pp. 1025–1029, 2013.

[28] D. Papkov, A. Goponenko, O. C. Compton, Z. An, A. Moravsky, X. Z. Li, S. T. Nguyen, and Y. A. Dzenis, Improved graphitic structure of continuous carbon nanofibers via graphene oxide templating, *Adv. Funct. Mater.*, vol. 23, pp. 5763–5770, 2013.

[29] Q. Wang, Y. Du, Q. Feng, F. Huang, K. Lu, J. Liu, and Q. Wei, Nanostructures and surface nanomechanical properties of polyacrylonitrile/graphene oxide composite nanofibers by electrospinning, *J. Appl. Polym. Sci.*, vol. 128, no. 2, pp. 1152–1157, 2013.

[30] S. Thampi, V. Muthuvijayan, and R. Parameswaran, Mechanical characterization of high-performance graphene oxide incorporated aligned fibroporous poly(carbonate urethane) membrane for potential biomedical applications, *J. Appl. Polym. Sci.*, vol. 132, no. 16, p. n/a–n/a, 2015.

[31] C. Wang, Y. Li, G. Ding, X. Xie, and M. Jiang, Preparation and characterization of graphene oxide/poly(vinyl alcohol) composite nanofibers via electrospinning, *J. Appl. Polym. Sci.*, vol. 127, no. 4, pp. 3026–3032, 2013.

[32] O. J. Yoon, C. Y. Jung, I. Y. Sohn, H. J. Kim, B. Hong, M. S. Jhon, and N. E. Lee, Nanocomposite nanofibers of poly(D, L-lactic-co-glycolic acid) and graphene oxide nanosheets, *Compos. Part A Appl. Sci. Manuf.*, vol. 42, no. 12, pp. 1978–1984, 2011.

[33] Y.-H. Yu, C.-C. Chan, Y.-C. Lai, Y.-Y. Lin, Y.-C. Huang, W.-F. Chi, C.-W. Kuo, H.-M. Lin, and P.-C. Chen, Biocompatible electrospinning poly(vinyl alcohol) nanofibres embedded with graphene-based derivatives with enhanced conductivity, mechanical strength and thermal stability, *RSC Adv.*, vol. 4, no. 99, pp. 56373–56384, 2014.

[34] H. R. Pant, C. H. Park, L. D. Tijing, A. Amarjargal, D.-H. Lee, and C. S. Kim, Bimodal fiber diameter distributed graphene oxide/nylon-6 composite nanofibrous mats via electrospinning, *Colloids Surfaces A Physicochem. Eng. Asp.*, vol. 407, pp. 121–125, 2012.

[35] Y. Y. Qi, Z. X. Tai, D. F. Sun, J. T. Chen, H. B. Ma, X. B. Yan, B. Liu, and Q. J. Xue, Fabrication and characterization of poly(vinyl alcohol)/graphene oxide nanofibrous biocomposite scaffolds, *J. Appl. Polym. Sci.*, vol. 127, no. 3, pp. 1885–1894, 2013.

[36] L. Jin, D. Yue, Z.-W. Xu, G. Liang, Y. Zhang, J.-F. Zhang, X. Zhang, and Z. Wang, Fabrication, mechanical properties, and biocompatibility of reduced graphene oxide-reinforced nanofiber mats, *RSC Adv.*, vol. 4, p. 35035, 2014.

[37] Y. Liu, M. Park, H. Kyoung, B. Pant, J. Choi, Y. Wan, J. Youb, S. Park, and H. Kim, Facile preparation and characterization of poly (vinyl alcohol)/chitosan/graphene oxide biocomposite nanofibers, *J. Ind. Eng. Chem.*, vol. 20, no. 6, pp. 4415–4420, 2014.

[38] X. An, H. Ma, B. Liu, and J. Wang, Graphene oxide reinforced polylactic acid/polyurethane antibacterial composites, *J. Nanomater.*, vol. 2013, pp. 1–7, 2013.

[39] H. Ma, W. Su, Z. Tai, D. Sun, X. Yan, B. Liu, and Q. Xue, Preparation and cytocompatibility of polylactic acid/hydroxyapatite/graphene oxide nanocomposite fibrous membrane, *Chinese Sci. Bull.*, vol. 57, no. 23, pp. 3051–3058, 2012.

[40] Y. C. Shin, J. H. Lee, L. Jin, M. J. Kim, Y.-J. Kim, J. K. Hyun, T.-G. Jung, S. W. Hong, and D.-W. Han, Stimulated myoblast differentiation on graphene oxide-impregnated PLGA-collagen hybrid fibre matrices, *J. Nanobiotechnology*, vol. 13, no. 1, p. 21, 2015.

[41] C. Zhang, L. Wang, T. Zhai, X. Wang, Y. Dan, and L.-S. Turng, The surface grafting of graphene oxide with poly(ethylene glycol) as a reinforcement for poly(lactic acid) nanocomposite scaffolds for potential tissue engineering applications, *J. Mech. Behav. Biomed. Mater.*, vol. 53, pp. 403–413, 2016.

[42] J. Song, H. Gao, G. Zhu, X. Cao, X. Shi, and Y. Wang, The preparation and characterization of polycaprolactone/graphene oxide biocomposite nano fiber scaffolds and their application for directing cell behaviors, *Carbon N. Y.*, vol. 95, pp. 1039–1050, 2015.

[43] X. Jing, H.-Y. Mi, M. R. Salick, T. M. Cordie, X.-F. Peng, and L.-S. Turng, Electrospinning thermoplastic polyurethane/graphene oxide scaffolds for small diameter vascular graft applications., *Mater. Sci. Eng. C. Mater. Biol. Appl.*, vol. 49, pp. 40–50, 2015.

[44] B. Ardeshirzadeh, N. A. Anaraki, M. Irani, L. R. Rad, and S. Shamshiri, Controlled release of doxorubicin from electrospun PEO/chitosan/graphene oxide nanocomposite nanofibrous scaffolds., *Mater. Sci. Eng. C. Mater. Biol. Appl.*, vol. 48, pp. 384–90, 2015.

[45] Y. Luo, H. Shen, Y. Fang, Y. Cao, J. Huang, M. Zhang, J. Dai, X. Shi, and Z. Zhang, "Enhanced proliferation and osteogenic differentiation of mesenchymal stem cells on graphene oxide-incorporated electrospun poly(lactic-co-glycolic acid) nanofibrous mats., *ACS Appl. Mater. Interfaces*, vol. 7, no. 11, pp. 6331–9, 2015.

[46] R. Ding, Z. Luo, X. Ma, X. Fan, L. Xue, X. Lin, and S. Chen, High sensitive sensor fabricated by reduced graphene oxide/polyvinyl butyral nanofibers for detecting Cu (II) in water, *Int. J. Anal. Chem.*, vol. 2015.

[47] X. Su, J. Ren, X. Meng, X. Ren, and F. Tang, A novel platform for enhanced biosensing based on the synergy effects of electrospun polymer nanofibers and graphene oxides., *Analyst*, vol. 138, no. 5, pp. 1459–66, 2013.

[48] Z. L. Xu, B. Zhang, and J. K. Kim, Electrospun carbon nanofiber anodes containing monodispersed Si nanoparticles and graphene oxide with exceptional high rate capacities, *Nano Energy*, vol. 6, pp. 27–35, 2014.

[49] Shilpa, B. M. Basavaraja, S. B. Majumder, and A. Sharma, "Lectrospun hollow glassy carbon—reduced graphene oxide nano fibers with encapsulated ZnO nanoparticles: a free standing anode for Li-ion, *J. Mater. Chem. A Mater. energy Sustain.*, vol. 3, pp. 1–8, 2015.

[50] B.-H. Kim and K. S. Yang, Structure and electrochemical properties of electrospun carbon fiber composites containing graphene, *J. Ind. Eng. Chem.*, vol. 20, no. 5, pp. 3474–3479, 2014.

[51] H. Ke, Z. Pang, Y. Xu, X. Chen, J. Fu, Y. Cai, F. Huang, and Q. Wei, Graphene oxide improved thermal and mechanical properties of electrospun methyl stearate/polyacrylonitrile form-stable phase change composite nanofibers, *J Therm Anal Calorim*, vol. 117, pp. 109–122, 2014.

[52] C. Menchaca-Campos, C. García-Pérez, I. Castañeda, M. A. García-Sánchez, R. Guardián, and J. Uruchurtu, Nylon/graphene oxide electrospun composite coating, *Int. J. Polym. Sci.*, vol. 2013, pp. 1–9, 2013.

[53] H. R. Pant, P. Pokharel, M. K. Joshi, S. Adhikari, H. J. Kim, C. H. Park, and C. S. Kim, Processing and characterization of electrospun graphene oxide/polyurethane composite nanofibers for stent coating, *Chem. Eng. J.*, vol. 270, pp. 336–342, 2015.

[54] J. Zhao, N. Dehbari, W. Han, L. Huang, and Y. Tang, Electrospun multi-scale hybrid nano fiber/net with enhanced water swelling ability in rubber composites, *Mater. Des.*, vol. 86, pp. 14–21, 2015.

[55] L. Tan, L. Gan, J. Hu, Y. Zhu, and J. Han, Functional shape memory composite nanofibers with graphene oxide filler, *Compos. Part A Appl. Sci. Manuf.*, vol. 76, pp. 115–123, 2015.

[56] Q. Bao, H. Zhang, J. Yang, S. Wang, D. Y. Tang, R. Jose, S. Ramakrishna, C. T. Lim, and K. P. Loh, Graphene-polymer nanofiber membrane for ultrafast photonics, *Adv. Funct. Mater.*, vol. 20, no. 5, pp. 782–791, 2010.

[57] H. R. Pant, B. Pant, P. Pokharel, H. J. Kim, L. D. Tijing, C. H. Park, D. S. Lee, H. Y. Kim, and C. S. Kim, Photocatalytic Ti02-RGO/nylon-6 spider-wave-like nano-nets via electrospinning and hydrothermal treatment, *J. Memb. Sci.*, vol. 429, pp. 225–234, 2013.

Section 2

Carbon Nanotubes and Diamonds

Chapter 6

Flexible Low-Voltage Carbon Nanotube Heaters and their Applications

Seyram Gbordzoe, Rachit Malik, Noe Alvarez,
Robert Wolf and Vesselin Shanov

Additional information is available at the end of the chapter

http://dx.doi.org/10.5772/64054

Abstract

Carbon nanotube heaters recently gained more attention due to their efficiency and relative ease of fabrication. In this chapter, we report on the design and fabrication of low-voltage carbon nanotube (CNT) heaters and their potential applications. CNT sheets drawn from CNT arrays have been used to make the heaters. The sheet resistance of the CNT sheet is dependent on the number of layers accumulated during their formation, and it ranges from 3.57 kΩ/sq. for a 1-layer sheet to 6.03 Ω/sq. for a 300-layer sheet. The fabricated and studied CNT heaters revealed fast heating and cooling rate. Potential applications of these heating devices have been illustrated by manufacturing and testing heatable gloves and via deicing experiments using low-voltage CNT heaters.

Keywords: CNT sheet, CNT heater, heated gloves

1. Introduction

Lightweight and low-power-consuming heaters that can be easily integrated in different materials and devices are needed for many industries including aerospace and medicine. The cold seasons usually bring the risk for several health problems, such as hypothermia, frostbite, flu, and even heart attacks. These adverse conditions make it difficult for people to conduct their daily routines effectively and safely during cold weather, due to the lack of proper gear to protect them. Some of the main attributes of a heating material, which can be safely incorporated into apparel, are lightweight, flexibility, fast heating rate, and energy efficiency. CNT sheets meet most of these qualities and thus are potential candidates as heating materials.

Carbon nanotubes have been the subject of immense interest since their official discovery in 1991 [1]. They attracted much attention due to their unique electrical, structural, mechanical and thermal conductivities, and their multiple potential applications [2–5]. It has been shown that CNT arrays can be processed into yarns and sheets, which have excellent properties [6–11]. One of the newer applications of CNT sheets is their use as a heating material [12–17]. The resistivity of CNT films can be controlled by the number of layers within the sheet or by the reagents used for their densification [17, 18]. CNTs can be incorporated into polymers to form composites for heating applications [16, 19]. However, CNT/polymer composites have a limited range of performance temperatures due to possible decomposition of the polymers [20]. Powdered CNTs dispersed in solvents have been used to coat cotton fabric and create CNT/textile heaters [21]. CNT-based heaters have also found applications as efficient and durable electrodes for Faradaic water splitting [22]. CNT sheet heaters have been fabricated as transparent film heaters, due to their high transmittance and good heating capabilities [12, 15–17]. CNT heaters have also demonstrated to be more efficient than commercial heating materials such as nichrome [18, 23]. Additionally, CNT-based heaters revealed rapid heating capabilities [23], which make them ideal candidates for applications in aerospace and heated clothing.

In this study, we fabricated CNT sheets from CNT arrays produced by the chemical vapor deposition (CVD) process. These sheets were densified by solvents and used for the manufacture of CNT heaters. The goal of this work is to study how the heater's design (number of layers in the CNT sheets, their post-processing) influences its performance, including the voltage required to generate heat and the rapid heating response. Finally, some possible applications of the CNT heater are provided in this study.

2. Experimental methods

CNT sheet fabrication started with the synthesis of vertically aligned, spinnable CNTs, typically about 400 μm in length. Thin films of Fe and Co were used as a catalyst (1.2 nm), which were sputtered on 4-inch Si wafers with 5-nm Al_2O_3 as a buffer sublayer. The obtained surface-engineered substrates were placed in a deposition reactor ET 3000 made by chemical vapor deposition (CVD) Equipment Corporation for the growth of the CNT arrays. Details of the growth process have been published elsewhere [24].

Upon completion of the CNT growth, detached, drawable CNT arrays were transferred onto an Si wafer or glass. CNT sheets with the desired number of layers were then assembled by continuously accumulating CNT ribbon onto a drum covered with a Teflon film as shown in **Figure 1**.

The resulting CNT sheets were densified, layer by layer, using acetone, as they were collected onto the drum. This solvent was used because a previous work showed that it can significantly improve the electro-thermal behavior of CNT sheets [18]. Each revolution of the Teflon drum added one CNT layer of 50–70 nm thick ribbon onto the sheet. CNT sheets with various numbers of layers (1, 10, 100, 200 and 300) were produced by this technique.

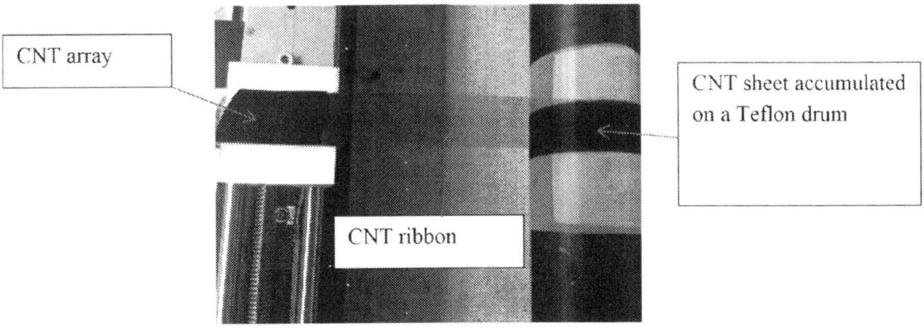

Figure 1. CNT ribbons being collected on a Teflon-covered drum from a CNT array.

CNT sheets were transferred onto a polyethylene terephthalate (PET) tape and then cut with a laser machine (Micro Machining Systems by Oxford Lasers, wavelength of 532 nm). PET tape was used as a supporting substrate, because this material is thermally stable up to 200°C and makes handling the CNT heaters easier. Samples were cut into 10 mm × 60 mm dimensions with a laser power of 5% and at a speed of 1 mm/s. The laser machine was also used to cut copper foils into 20 mm × 5 mm strips at a laser power of 40% and speed of 1 mm/s. The copper strips were attached at both ends of the CNT sheet with silver paste as shown in **Figure 2**, to form the CNT heater.

Figure 2. CNT sheet assembled between two polyethylene terephthalate (PET) tapes with Cu electrodes attached at both ends to form a CNT heater.

A DC power source (Hewlett Packard E 3612A) was used to supply power to the CNT heaters. The surface temperatures of the CNT heaters were measured using an infrared (IR) camera (FLIR T640). The average temperature was determined by measuring the surface temperature at three different locations across the sheet. The effective area of the CNT sheet actually heated up was 10 mm × 50 mm for all samples. The voltage was stepped until pre-determined heater temperatures were reached, and the corresponding current was measured. Measurements were taken at room temperature and pressure, after the heaters were allowed to reach an equilibrium temperature.

The sheet resistance of the CNT sheets was measured using a Jandel 4-probe instrument (Model RM3000). Prior to the measurement of the sheet resistance, samples were placed on a flat substrate, and then the sheet resistance was measured in the direction parallel and perpendicular to the alignment of the CNT sheet.

3. Results and discussion

3.1. Effect of the number of layers on sheet resistance

The sheet resistance of samples was measured in both parallel and perpendicular to the drawing directions, using four probe techniques, as shown in **Figure 3**. This method involves passing current through the two outer probes, while the resulting voltage drop is measured by the two inner probes. The measured sheet resistance is presented in **Table 1**.

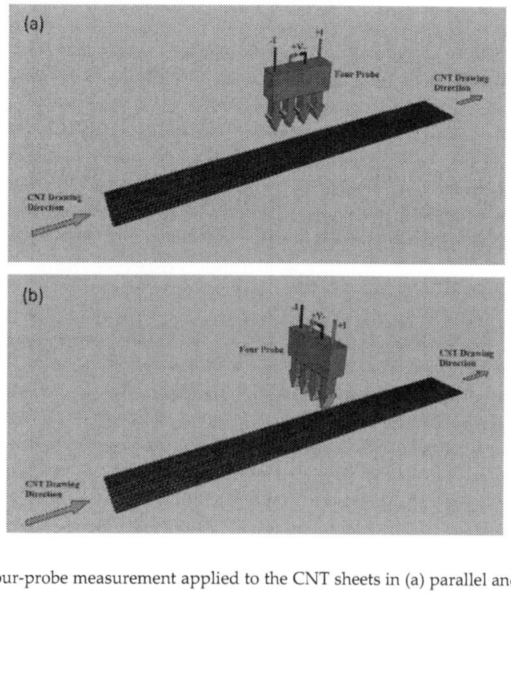

Figure 3. Scheme of the four-probe measurement applied to the CNT sheets in (a) parallel and (b) perpendicular to the drawing direction.

http://dx.doi.org/10.5772/64054

Number of layers	Parallel sheet resistance (Ω/sq.)	Perpendicular sheet resistance (Ω/sq.)
1	3567.80 ± 1413.03	3463.40 ± 727.35
10	171.26 ± 13.13	187.08 ± 21.72
100	17.21 ± 0.52	18.10 ± 0.39
200	9.89 ± 0.52	11.24 ± 0.69
300	6.03 ± 0.23	6.51 ± 0.18

Table 1. Resistance of CNT sheet in parallel and perpendicular to the drawing direction.

As the number of layers increased, the sheet resistance decreased accordingly. A growth in the number of layers leads to an increase in the density of CNTs, as shown by SEM images in **Figure 4**. The increase in density leads to a reduction in the electrical resistance, because the tube-to-tube contacts were increased, allowing for more electrons to be transferred through the sample.

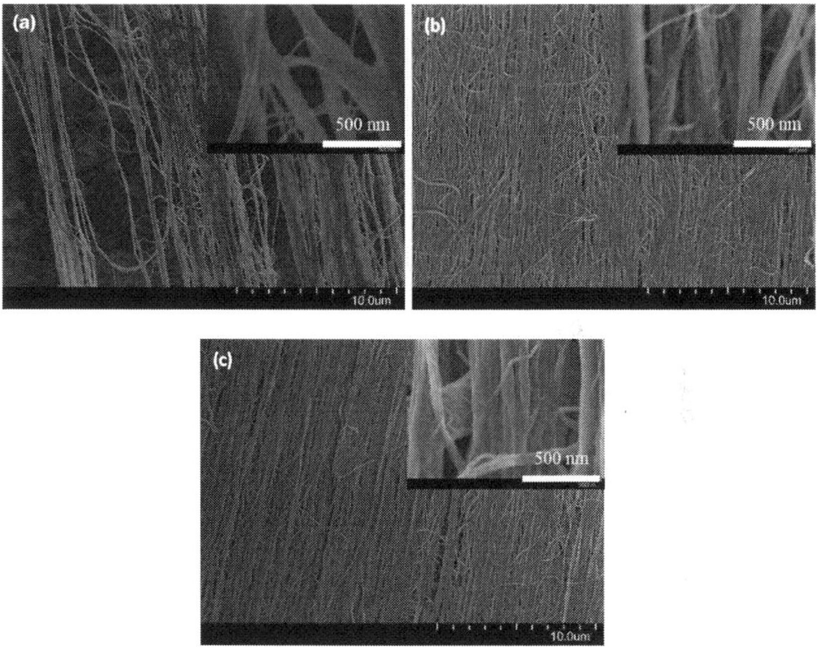

Figure 4. SEM images of (a) 1-, (b) 10- and (c) 100-layer-densified CNT sheets (high magnification in inset) showing an increase in density with increasing number of layers.

The parallel sheet resistance is lower than the perpendicular sheet resistance for 10, 100, 200 and 300 layers. This is because there is less resistance when electrons travel along the tube length, compared to the lateral tube-to-tube interactions, which take place in the perpendicular case. This is, however, not the case for a 1-layer sheet. As shown in **Figure 4a**, the probability of the 4 probes hitting the same bundle of sheets, to measure the resistance, is lower in the perpendicular direction compared to the parallel direction. The anisotropy of the sheets is, however, lower than what has been in reported in some cases. Inoue et al. reported an anisotropy of 7.3 between the sheet resistances in the parallel direction compared to that in the perpendicular direction [25].

3.2. Electrical properties of CNT heaters

The CNT sheets were connected to a DC power source, and the corresponding temperature was measured using an IR camera in ambient environment, as shown in **Figure 5**. The heat generated across the sheet is as a result of the Joule heating phenomenon. Joule heating is a phenomenon that occurs when electrical current is passed through a material with an electrical resistance. The resistance inherent to the material leads to a conversion of electrical energy to thermal energy. This is caused by the collision of the moving electrons with the atoms that are the main material's constituents. The quantity of heat (Q) that is yielded from the applied electrical energy is proportional to the square of the current (I) multiplied by the resistance (R) over a period of time (t) as per Eq. (1).

$$Q = I^2 Rt \tag{1}$$

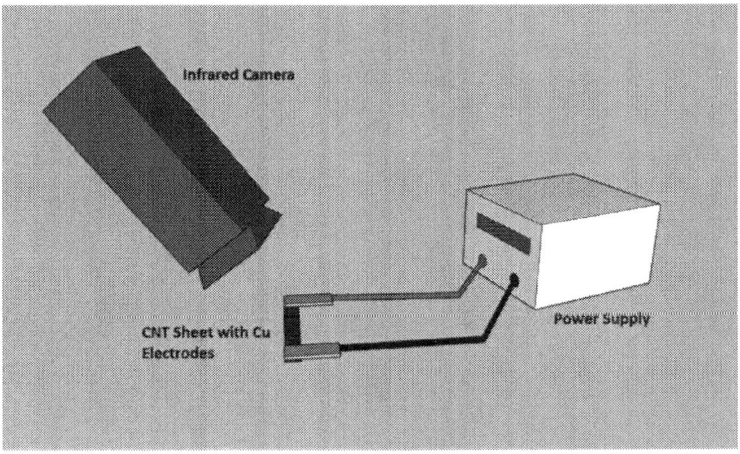

Figure 5. Schematic illustration of the temperature acquisition setup using an infrared (IR) camera.

However, we usually use another parameter, known as power (P), to describe how much electrical energy is converted into thermal energy. This is simply the quantity of heat (Q) divided by the time (t) as per Eq. (1), which leads to Eq. (2).

$$P = I^2 R = \frac{V^2}{R} \tag{2}$$

The power (P) generated by the CNT sheet when connected to a voltage source is also directly proportional to the voltage (V) across the sheet and the current (I) passing through the sheet (Eq. (2)). Different CNT heaters were connected to a DC power source, and the voltage was changed to correspond to temperatures 30, 50, 70, 90 and 110°C. The resultant current and voltage at each temperature were noted and used to calculate the corresponding power. The effect of the number of layers on the power needed to reach a particular temperature is illustrated in **Figure 6**. It can be seen that, as the number of layers increased, the power needed to generate a particular temperature also increased. This can be explained using Eq. (2), which shows that the resistance is inversely proportional to the power. An increase in the number of layers leads to a decrease in resistance, and this in turn leads to an increase in power needed to attain a particular temperature. Therefore, sheets with high resistance need a lower power to attain the same temperature as sheets with lower resistances.

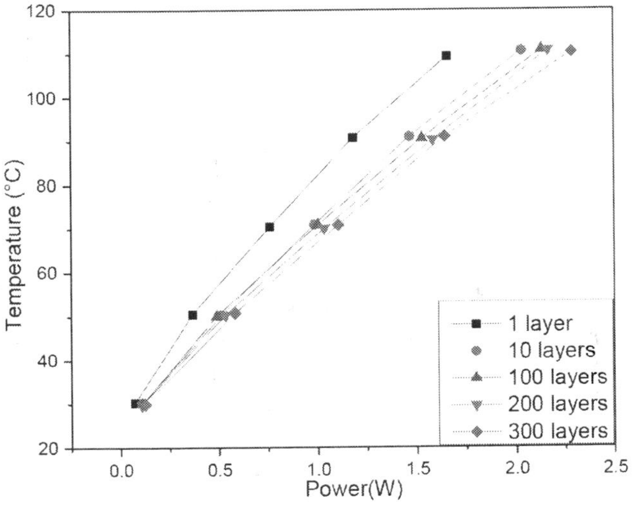

Figure 6. Power versus temperature for different number of layers within the CNT sheet.

Equation 2 also shows that the higher the resistance of the sheet, the greater the voltage needed to attain a specific temperature. This trend is clearly seen in **Figure 7**. The lower number layer sheets (1 and 10) needed very high voltages to attain the same temperature as the high number layer sheets (100, 200 and 300). For practical applications, the heater might be limited by the available low voltage supply. In such a case, CNT sheets with less number of layers would not be a feasible option, and higher number of layer sheets would be preferred.

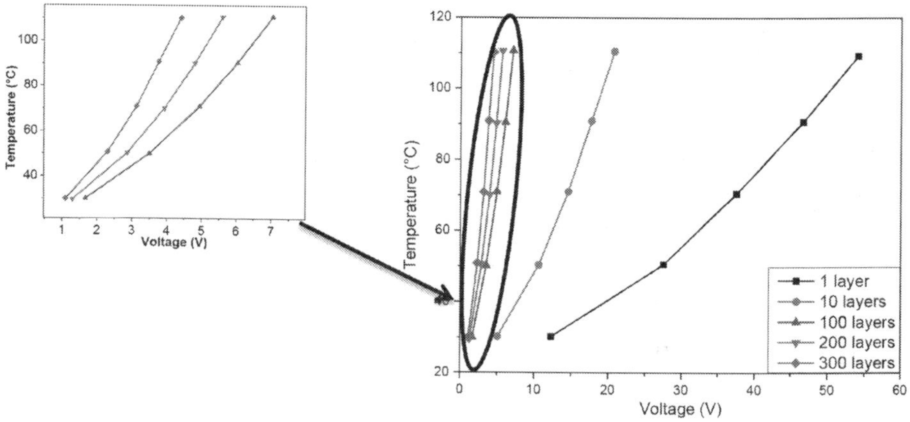

Figure 7. Voltage versus temperature for different number of layers within the CNT sheet.

The electrical resistivity of the CNT sheets was also investigated in this study. The resistivity (ρ) is related to the sheet resistance (R_s) and thickness (t) of the sheet by Eq. (3).

$$\rho = \frac{R_S}{t} \tag{3}$$

The resistivity of the CNT sheet can therefore be calculated if the thickness of the CNT sheet is known. Cross-sectional SEM images were taken on a selected number of samples to determine their thicknesses. The sheets were cut perpendicular to the drawing direction with a laser powered 0.5%, using a cutting speed of 0.1 mm/s.

The lower power and reduced speed of the laser were important to ensure a clean cut of the sheet and avoid pullouts of the CNTs disturbing the thickness uniformity. In order to increase the conductivity of the CNT sheets for SEM imaging, the samples were mounted on a vertical stub and coated with a thin layer of Au/Pd by a sputtering machine (Denton Vacuum Desk IV) using a sputtering time of 25 s. SEM images of the different sheet thicknesses are displayed in **Figure 8**. The SEM images show the CNT samples with uniform thickness across the sheet. The average thickness of 100, 200 and 300 layer sheets was 6.25, 13.32 and 20.31 μm, respectively. The calculated resistivity in the parallel direction using Eq. (2) for 100, 200 and 300 layer sheets was 1.08×10^{-2} Ω cm, 1.32×10^{-2} Ω cm, and 1.23×10^{-2} Ω cm, respectively. The resistivity of the sheets is comparable to the data reported in the literature where MWCNT films have been studied revealing a resistivity of 20×10^{-3} Ω cm [26]. Our resistivity is higher than those sheets composed of single-wall carbon nanotubes (SWCNTs), reported to be in the range of 10^{-4} to 10^{-5} Ω cm [4]. The high resistivity in our work is due to several factors, including the packing density of the nanotubes and the nature of the MWCNTs, which are less conductive than SWCNTs.

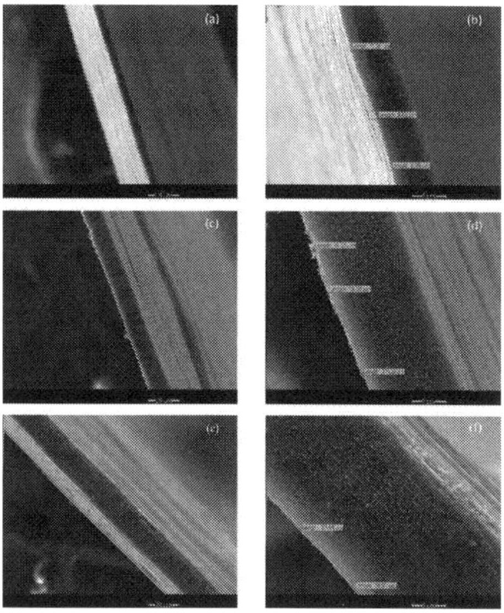

Figure 8. Cross-sectional SEM images showing the thickness of: (a) and (b) 100-layer CNT sheet; (c) and (d) 200-layer CNT sheet; and (e) and (f) 300-layer CNT sheet.

It has been shown in the literature that an improved packing density (by stretching and pressing) leads to improvement in electrical and mechanical properties [7, 27]. In the future, we intend to use these two approaches in order to increase the density of our CNT sheets and decrease their resistivity.

3.3. Temperature profile of CNT sheets

The heat generated by the CNT heater, when connected to a power source, is lost into the ambient environment during heating. The total heat is dissipated into the air and substrate through three mechanisms: conduction, convection and radiation [12].

The heating profiles with time of CNT sheets at selected low voltages (1.5, 3 and 4.5 V) are presented in **Figure 9**. The obtained data mimic possible low-voltage applications of the heaters. Sheets with 1 and 10 layers have not been included because they show negligible increase in temperature at the applied voltages.

The observed temperature evolution is divided into 3 stages: heating, saturation, and cooling. Cooling happens after the voltage source is turned off. During the saturation stage, the heat gained by the applied electric power is almost equal to the heat lost by radiation and convection. The average time for the heater to reach the temperature saturation is 20 s. However, the average cooling time depends on the final temperature of the heater and, therefore, varies, as shown in **Figure 9**. This fast heating time is a desirable characteristic of a heater that will be incorporated into clothing or textiles, to keep the body warm during cold seasons. This quality is also beneficial for deicing applications. We present some feasible applications of the CNT heaters created and studied in this work.

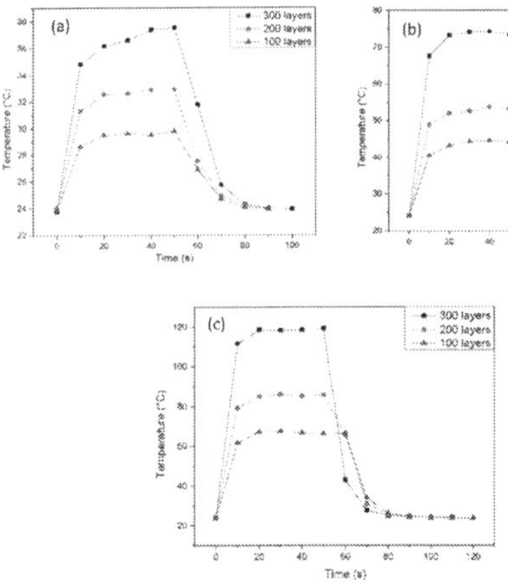

Figure 9. Heating profile as a function of time for 100-, 200- and 300-layer CNT sheets at (a) 1.5 V, (b) 3 V, and (c) 4.5 V.

4. Applications of CNT heaters

4.1. Heatable glove

CNT heaters are easy to design and manufacture, which makes them attractive for a variety of applications. Their lightweight, mechanical flexibility and low voltage make the CNT heaters appealing candidates for apparel, to protect operators performing in cold weather. We have fabricated CNT heaters and embedded them into gloves, based on the approach described in this work. The heaters were made using 200 and 150 layer sheets. Such gloves are lightweight since the CNT heaters do not add significant mass (total change of 1.21 g) and can also be operated with either a 9-V battery or a rechargeable Li-ion (2700 mAh, 3.7 V) battery, as shown in **Figure 10**. This type of application has advantages because of the ease of fabrication of the CNT heaters and the low voltage needed to power them.

4.2. Deicing applications

The use of CNT heaters for deicing applications was also studied. This approach was similar to that described by Janas et al. [23]. A 300-layer CNT heater was attached to the right wing of a model airplane. The test model was inserted into a freezer with a continuous voltage (3V) supply attached to the heater. Droplets of water were then sprinkled on both wings of the model airplane, shown in **Figure 11a**, and it was left in the freezer. The thermal image before ice formation is shown in **Figure 11b**, where the right wing temperature is above 0°C. After an hour of cooling, the wing without the CNT heater had visible ice formation on its surface, while the wing with the functioning heater was dry as shown in **Figure 11c**. The corresponding infrared image, after ice formation, is shown in **11d**, where the entire model airplane is frozen except for the right wing. This experiment shows that CNT heaters can be used in deicing applications to prevent the formation of ice on airplanes, which is a major issue in the aerospace industry. The heater can be incorporated into the surface of the fuselage, which is made of polymer composite for most of the military and some of the advanced commercial airplanes.

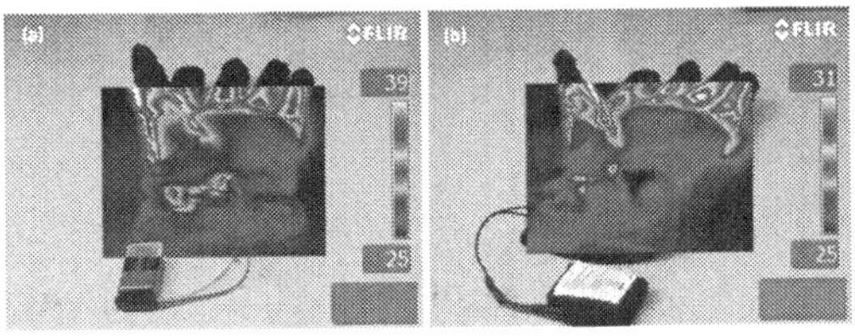

Figure 10. CNT-heated gloves connected to (a) 9-V battery and (b) rechargeable lithium ion battery. The color scale represents temperature in °C obtained by an IR camera.

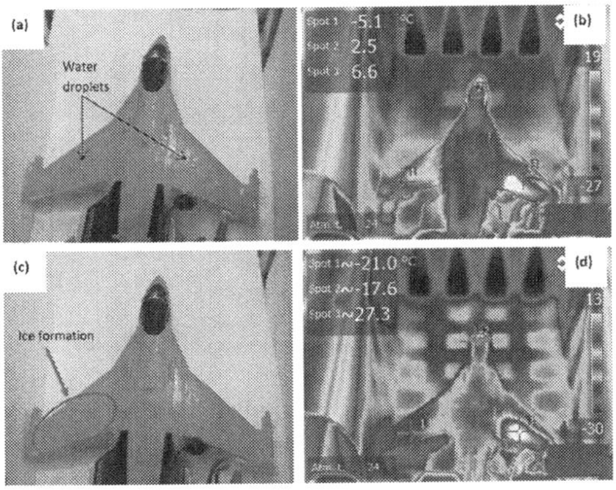

Figure 11. (a) Droplets of water on the wings of a model airplane with related (b) IR thermal image; (c) ice formation on the left wing of the model airplane after an hour of freezing and related (d) IR thermal image.

5. Conclusions

A simple and easy way for fabricating flexible and low-voltage CNT heaters was described. The resistance of the CNT sheets with different numbers of layers was studied. It was observed that as the number of layers increased, the sheet resistance decreased, showing that this property is dependent on the thickness of the sheet. The heating and cooling rates of these heaters were also studied. The obtained results suggested that the fast heating rate of the heaters makes them attractive for manufacturing body protection clothing used in cold weather. An example of the latter was demonstrated in this work by incorporating CNT heaters into a glove. Maximum glove temperatures of 39°C and 31°C were attained using a 9-V battery and Li-ion battery, respectively. Another application demonstrated that the CNT heater can find application in deicing solid surfaces exposed to freezing weather, such as airplanes. The CNT sheet heaters can be easily incorporated in polymer composites recently used as materials of choice for making advanced cars and fuselages.

Acknowledgements

This work was funded by the National Science Foundation (NSF) through the following grants: CMMI-0727250, SNM-1120382, ERC-0812348, and DURIP-ONR. The partial support by the National Institute for Occupational Safety and Health Pilot Research Project Training Program of the UC Education and Research Center Grant #T42/OH008432-10 is also acknowledged. The authors are grateful for the financial support from the UC Office of Research and the Mathewson Renewable Energy Research Fund. Seyram Gbordzoe is thankful for the graduate assistantship provided to him by the College of Engineering and Applied Science (CEAS) at UC and the Cincinnati Engineering Enhanced Mathematics and Science (CEEMS) program (NSF, DRL-1102990).

Author details

Seyram Gbordzoe[1], Rachit Malik[1], Noe Alvarez[2], Robert Wolf[3] and Vesselin Shanov[1,2*]

*Address all correspondence to: vesselin.shanov@uc.edu

1 Department of Mechanical and Materials Science Engineering, University of Cincinnati, Cincinnati, OH, USA

2 Department of Biomedical, Chemical and Environmental Engineering, University of Cincinnati, Cincinnati, OH, USA

3 Department of Electrical Engineering, University of Cincinnati, Cincinnati, OH, USA

References

[1] Iijima, S., *Helical microtubules of graphitic carbon*. Nature, 1991. 354(6348): pp. 56–58.

[2] Jung , D., M. Han, L.J. Overzet, and G.S. Lee, *Effect of hydrogen pretreatment on the spin-capability of a multiwalled carbon nanotube forest.* Journal of Vacuum Science & Technology B, 2013. 31(6): p. 06FI02.

[3] Treacy, M.M.J., T.W. Ebbesen, and J.M. Gibson, *Exceptionally high Young's modulus observed for individual carbon nanotubes.* Nature, 1996. 381(6584): pp. 678–680.

[4] Ebbesen, T.W., H.J. Lezec, H. Hiura, J.W. Bennett, H.F. Ghaemi, and T. Thio, *Electrical conductivity of individual carbon nanotubes.* Nature, 1996. 382(6586): pp. 54–56.

[5] Jung, D., M. Han, and G.S. Lee, *Gas sensor using a multi-walled carbon nanotube sheet to detect hydrogen molecules.* Sensors and Actuators A: Physical, 2014. 211: pp. 51–54.

[6] Pöhls, J.-H., M.B. Johnson, M.A. White, R. Malik, B. Ruff, C. Jayasinghe, M.J. Schulz, and V. Shanov, *Physical properties of carbon nanotube sheets drawn from nanotube arrays.* Carbon, 2012. 50(11): pp. 4175–4183.

[7] Liu, Q., M. Li, Y. Gu, Y. Zhang, S. Wang, Q. Li, and Z. Zhang, *Highly aligned dense carbon nanotube sheets induced by multiple stretching and pressing.* Nanoscale, 2014. 6(8): pp. 4338–4344.

[8] Zhang, M., S. Fang, A.A. Zakhidov, S.B. Lee, A.E. Aliev, C.D. Williams, K.R. Atkinson, and R.H. Baughman, *Strong, transparent, multifunctional, carbon nanotube sheets.* Science, 2005. 309(5738): pp. 1215–1219.

[9] Chun, K.-Y., et al., *Hybrid carbon nanotube yarn artificial muscle inspired by spider dragline silk.* Nat Commun, 2014. 5(3322).

[10] Zhang, M., K.R. Atkinson, and R.H. Baughman, *Multifunctional carbon nanotube yarns by downsizing an ancient technology.* Science, 2004. 306(5700): pp. 1358–1361.

[11] Jayasinghe, C., T. Amstutz, M.J. Schulz, and V. Shanov, *Improved processing of carbon nanotube yarn.* Journal of Nanomaterials, 2013. 2013: p. 7.

[12] Jung, D., M. Han, and G.S. Lee, *Flexible transparent conductive heater using multiwalled carbon nanotube sheet.* Journal of Vacuum Science & Technology B, 2014. 32(4): p. 04E105.

[13] Wu, Z., Q. Xu, J. Wang, and J. Ma, *Preparation of large area double-walled carbon nanotube macro-films with self-cleaning properties.* Journal of Materials Science & Technology, 2010. 26(1): pp. 20–26.

[14] Jang, H.-S., S.K. Jeon, and S.H. Nahm, *The manufacture of a transparent film heater by spinning multi-walled carbon nanotubes.* Carbon, 2011. 49(1): pp. 111–116.

[15] Kim, D., H.-C. Lee, J.Y. Woo, and C.-S. Han, *Thermal behavior of transparent film heaters made of single-walled carbon nanotubes.* The Journal of Physical Chemistry C, 2010. 114(13): pp. 5817–5821.

[16] Kim, D., L. Zhu, D.-J. Jeong, K. Chun, Y.-Y. Bang, S.-R. Kim, J.-H. Kim, and S.-K. Oh, *Transparent flexible heater based on hybrid of carbon nanotubes and silver nanowires*. Carbon, 2013. 63: pp. 530–536.

[17] Jung, D., D. Kim, K.H. Lee, L.J. Overzet, and G.S. Lee, *Transparent film heaters using multi-walled carbon nanotube sheets*. Sensors and Actuators A: Physical, 2013. 199: pp. 176–180.

[18] Dawid Janas and Krzysztof K. Koziol, "Improved Performance of Ultra-Fast Carbon Nanotube Film Heaters," Jounal of Automation and Control Engineering, Vol. 2, No. 2, pp. 150–153 June, 2014. doi: 10.12720/joace.2.2.150–153.

[19] Kunmo, C., K. Dongouk, S. Yoonchul, L. Sangeui, M. Changyoul, and P. Sunghoon, *Electrical and thermal properties of carbon-nanotube composite for flexible electric heating-unit applications*. Electron Device Letters, IEEE, 2013. 34(5): pp. 668–670.

[20] Lee, E. and Y.G. Jeong, *Preparation and characterization of sulfonated poly(1,3,4-oxadiazole)/multi-walled carbon nanotube composite films with high performance in electric heating and thermal stability*. Composites Part B: Engineering, 2015. 77: pp. 162–168.

[21] Ilanchezhiyan, P., A.S. Zakirov, G.M. Kumar, S.U. Yuldashev, H.D. Cho, T.W. Kang, and A.T. Mamadalimov, *Highly efficient CNT functionalized cotton fabrics for flexible/wearable heating applications*. RSC Advances, 2015. 5(14): pp. 10697–10702.

[22] Janas, D., S. Kreft, S. Boncel, and K.K. Koziol, *Durability and surface chemistry of horizontally aligned CNT films as electrodes upon electrolysis of acidic aqueous solution*. Journal of Materials Science, 2014. 49(20): pp. 7231–7243.

[23] Janas, D. and K.K. Koziol, *Rapid electrothermal response of high-temperature carbon nanotube film heaters*. Carbon, 2013. 59: pp. 457–463.

[24] Alvarez, N.T., P. Miller, M. Haase, N. Kienzle, L. Zhang, M.J. Schulz, and V. Shanov, *Carbon nanotube assembly at near-industrial natural-fiber spinning rates*. Carbon, 2015. 86: pp. 350–357.

[25] Inoue, Y., Y. Suzuki, Y. Minami, J. Muramatsu, Y. Shimamura, K. Suzuki, A. Ghemes, M. Okada, S. Sakakibara, and H. Mimura, *Anisotropic carbon nanotube papers fabricated from multiwalled carbon nanotube webs*. Carbon, 2011. 49(7): pp. 2437–2443.

[26] Bacsa, W., A. Chatelain, T. Gerfin, R. Humphrey-Baker, L. Forro, and D. Ugarte, *Aligned carbon nanotube films: production and optical and electronic properties*. Science, 1995. 268(5212): pp. 845–847.

[27] Cheng, Q., J. Bao, J. Park, Z. Liang, C. Zhang, and B. Wang, *High mechanical performance composite conductor: multi-walled carbon nanotube sheet/bismaleimide nanocomposites*. Advanced Functional Materials, 2009. 19(20): pp. 3219–3225.

Chapter 7

Recent Trends of Reinforcement of Cement with Carbon Nanotubes and Fibers

Oxana V. Kharissova, Leticia M. Torres Martínez and Boris I. Kharisov

Additional information is available at the end of the chapter

http://dx.doi.org/10.5772/62292

Abstract

Recent achievements in the area of formation of carbon nanotubes (CNTs), nanocomposites, with cement are reviewed. The peculiarities of dispersion of CNTs in cementitious matrices are discussed, paying major attention to the CNT diameter, length and length-to-diameter ratio, concentration, functionalization, annealing, combination with other nanomaterials, and water-cement ratio. Several effects upon dispersion of carbon allotropes in concrete-water media are emphasized. It is also pointed out that the health impact should also be considered in further experiments on construction materials reinforced with CNTs.

Keywords: Carbon fibers, Carbon nanotubes, Cement, Composite, Dispersion, Reinforcement

1. Introduction

Concrete, steel, and asphalt coatings are conventional materials used in large scale and produced in huge quantities worldwide. In case of concrete (a mixture weak in tension and strong in compression), more than 2 tons per person are produced annually. Quantitatively, more than 11 billion metric tons are consumed every year all over the world. The cement industry is responsible for approximately 5–8% of all anthropogenic emissions of carbon dioxide worldwide [1]. Portland cement (containing, in dry phase, 63% calcium oxide, 20% silica, 6% alumina, and 3% iron (III) oxide, and small amounts of other substances) is a product with great, but not completely explored potential, despite intensive studies in the last century. In the last century, when the nanotechnology era began, at the end of the 1980s, novel nanomaterials started to be used as additives in numerous applications, in particular for reinforcement of construction materials, and certain attention was given to cement composites. Their complex structure, when studied at the nanolevel, will definitely lead to new generations of highly durable concretes with "smart" properties. In order to improve concrete properties, a series of nanomaterials can be applied, such as nanoengineered polymers, superplasticizers, high strength fibers, or silicon dioxide (**Figure 1**).

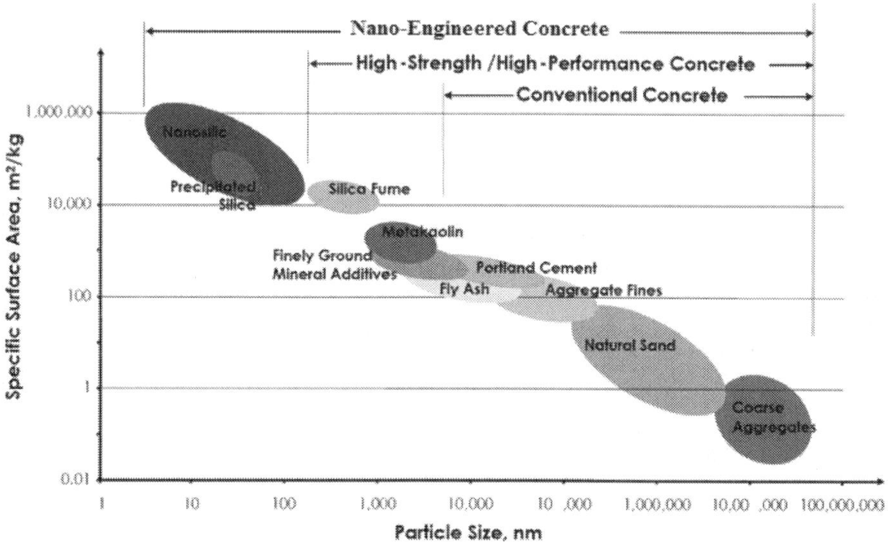

Figure 1. Particle size versus specific surface area scale in to concrete materials [2]. Reproduced with permission.

Carbon nanotubes (CNTs), well known for having extraordinary properties, are considered as major candidates for diverse applications in nanotechnology. CNTs have been incorporated into materials for cement production, with excellent results, upon low quantities of added material. Thus, reinforcing cement with low additions of multiwalled carbon nanotubes (MWCNTs), ranging from 0.05 to 0.5 wt%, can represent a remarkable enhancement in the mechanical properties of cement. It is expected that CNTs, when added to concrete, will increase compression strength beyond 200 MPa, thus allowing the construction of mile-high skyscrapers. Nowadays, CNTs and carbon nanofibers (CNF) are two of the most prospective advanced materials for application in cement-based products, for the construction industry, due to their excellent material properties. According to the reported results, CNTs can reduce the occurrence of cracks, decrease porosity, and improve mechanical properties, thus extending the cement durability. However, CNTs are insoluble in organic solvents and water, so surfactants should be normally added upon classic combination of ultrasonication and vigorous agitation, in order to disperse CNTs. There are other well-known less common specific methods for CNT "dissolution" in liquid media [3, 4]; some of which can be adjusted to concrete fabrication.

In this chapter, we discuss recent achievements of cement reinforcement with carbon allotropes. Taking into account a recently published excellent comprehensive review [5] in this area and other related publications [6, 7], we refer the particular aspects and peculiarities of dispersion of CNTs in cement mixtures.

2. Peculiarities of CNTs and CNFs dispersion in concrete

There are at least three main ways to add CNTs uniformly to concrete: addition to cement, to water, and as admixture. In the majority of reports, ultrasonic dispersion techniques have been widely used to uniformly disperse CNTs [8]. Ultrasonication can be applied alone after adding CNTs to the concrete mass, or together with surfactants. Generally, pristine CNTs can be directly mixed with cement, but their hydrophobicity limits solubilization in aqueous media. So, frequently the CNTs are functionalized with HNO_3/H_2SO_4 mixture, forming -OH and -COOH functionalized CNTs, before being added to cement [9]. Alternatively, carboxylic functionalization can be done upon ultrasonication and polyacrylic acid polymer addition [10]. The use of superplasticizers (see below), magnetic stirring [11], CNT growth onto cement particles [12], and dry mixing with cement [13] are other options to disperse CNTs and CNFs in concrete, frequently accompanied by ultrasonication [14].

Generally, two main strategies are applied to provide an interfacial bonding between cementitious matrix and CNTs: (a) formation of hydration products of the cement matrix and (b) formation of covalent bonds on CNT surface using functional groups. **Figure 2** shows a scheme of interaction of the C-S-H (this means calcium-silicate-hydrate) cement product and a -COOH (carboxylic group) on the CNT surface[15]. These effects of *surface functionalization* of MWCNTs with -COOH groups on the strength and structure of Portland cement composites have been deeply studied [16]. Thus, grafting of functional groups on the surface of the nanotubes allowed acceleration of cement hydration. The authors established that the use of carboxylated nanotubes contributed to early strength development. The MWCNT-reinforced composites are characterized by high content in calcium silicate hydrates and very dense structure. The maximal compressive strength of 64 MPa (20% increase over the reference material) was observed for the composite with 0.13% of MWCNTs (by the weight of cement). The addition of carboxylated MWCNTs, in the very low amount of 0.05%, provides a 30% increase of 1-day compressive strength of developed composites.

In addition to the CNT functionalization above, *surfactants* can be applied to decrease aggregation of CNTs, by modification of their surface, since CNTs tend to aggregate together in most solvents, due to van der Waals forces, and form nanotube clusters and bundles. CNTs need to be completely dispersed in concrete in order to explore their outstanding physical properties. Application of classic surfactants, such as sodium dodecyl sulfate (SDS) and Triton X-100, has been observed in a series of reports. The effect of surfactants on pressure-

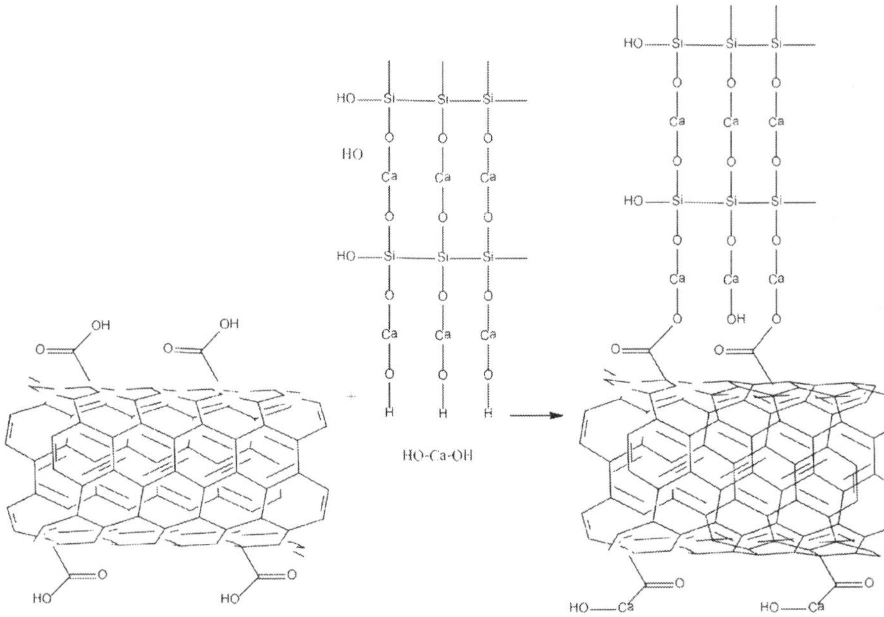

Figure 2. Reaction scheme between carboxylated nanotubes and cement hydration products ($Ca(OH)_2$ and C-S-H).

sensitivity of CNT-filled cement mortar composites has been studied using sodium dodecyl sulfate and sodium dodecylbenzene sulfonate (NaDDBS) as surfactants to disperse MWNTs-COOH (OD < 8 nm, ID 2–5 nm, length 10–30 μm) in cement mortar (**Figure 3**) [17]. The authors revealed that, in comparison with SDS-based composites, the composites with NaDDBS show a more stable and sensitive response of electrical resistance to external force. In this respect, NADDBS is a more efficient surfactant, in comparison with SDS, for dispersion of MWCNTs in cement.

Other nanotechnology classic surfactants are less common to use for CNT solubilization in concrete. Stable homogeneous suspensions of MWCNTs were prepared using gum arabic (GA) as dispersant (with an optimum GA concentration of 0.45 g/L^1) and were incorporated into Portland cement paste [18]. The authors revealed that the compressive strength, as well as the flexural strength of the Portland cement composite, can be considerably improved by adding the treated carbon nanotubes. At 0.08 wt% MWCNT concentration, the flexural strength reached an increase in 43.38%. MWCNTs act as bridges and networks across cracks and voids, which transfer the load in case of tension, and the interface bond strength between the

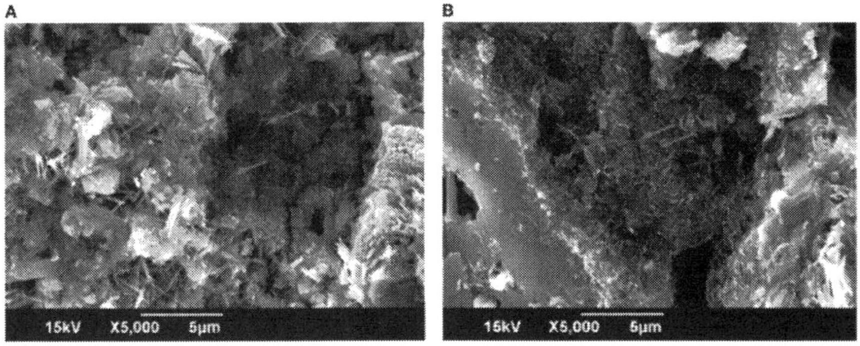

Figure 3. SEM images of CNT-filled cement mortar composites. (A) With NaDDBS and 1 wt% of MWNTs. (B) With SDS and 1 wt% of MWNTs. Reproduced with permission from [17].

nanotubes and matrix is very strong. The interfacial interactions between surface-modified nanotubes and hydrations (such as C-S-H and calcium hydroxide) of cement produce a high bonding strength, and increase the load-transfer efficiency from cement matrix to the reinforcement.

Among superplasticizers for concrete applications, we note polycarboxylate ether solution ("super plasticizer"), high-molecular-weight polyelectrolyte (i.e., poly (sodium 4-styrenesulfonate) or PSS) [19]. In a related report [20], the dispersity of MWCNTs in water and in cement paste, as well as its effect on the compressive strength of composition with MWCNTs and cement, was comparatively investigated using ultrasonic dispersion and three kinds of surface active agents: polycarboxylate superplasticizer, polyvinylpyrrolidone (PVP), and washing powder (WP). Agglomeration of MWCNTs takes place easily in cement paste and in water. The reinforce ability is poorly exhibited, even if the ultrasound is applied, in case of using, as a surface-active agent, a single polycarboxylate superplasticizer. WP and PVP are able to create MWCNT dispersions in cement paste and also in water under ultrasonic treatment; in addition, PVP enhances the compressive strength, while WP decreases it, as more air bubbles are brought inside the cement paste. In addition, several peculiarities exist on the dispersing carbon nanoparticles, in particular CNFs, in cement paste [21]. When a superplasticizer is used as surfactant, CNFs can be uniformly dispersed in water by ultrasonic processing. However, a uniform distribution of CNFs in cement paste will not be obtained by mixing water-superplasticizer-CNF dispersion with cement (used superplasticizer = high-range polycarboxylate-based water-reducing agent). To reach a better distribution in paste, either functionalized and highly dispersible CNFs should be used or the CNFs should initially be implanted or grown on cement particles. In addition, large cement particles prevent a uniform distribution, particularly when fibers are very small or are used in high doses. However, breaking cement grains into smaller particles is impractical and problematic. Fine-grain cement is very reactive and consumes a lot of water at a high rate. As a result, producing paste with typical water-cement ratios such as 0.40 or 0.45 is very difficult or even impossible. Therefore, the authors consider it important to use fresh cement with a minimal amount of large grains and clumps for making CNF-CNT reinforced composites.

3. Improvement of concrete properties by addition of CNTs and CNFs

It is well-known nowadays that mechanical and various other properties of concrete, indicated below, are considerably improved by addition of CNTs. Thus, the inclusion of MWCNTs (outer diameter > 50nm, length 10-20 mm) in the cement mix improves both the tensile fracture characteristics and compressive strength when not mixed with a surfactant compound [22]. The improvement in the mechanical properties in specimens with the addition of CNTs was observed more clearly with increasing curing age. The fracture mechanics test results indicated that the fracture properties of microconcrete and mortar are increased through proper dispersion of very low amount of MWCNTs (CNTs/c = 0.005). The vibration damping capacities of cement-based matrices with additions of MWCNTs (external diameter 20-40 nm, length 5-15 μm) were investigated with a free vibration testing method in an elastic system [23]. Adding small amounts of MWCNTs, several positive effects on the critical damping ratio (ζ) were observed for the cement matrix. This magnitude is increased in the MWCNT-cement composites because of frictions among MWCNT matrix and multiple intertubes.

As an example of improvement of other cement properties, MWCNT (external diameter 20-40 nm, length 5-15 μm) reinforced cement composites (MWFRCs) were prepared with surfactant dispersion, ultrasonic treatment, and subsequently high-speed shear mixing processes [24]. Several following parameters can be considerably improved by adding MWCNTs: critical crack mouth opening displacement (δ_C, up to 119.4%), flexural strength (σ_s, up to 54.8%) of the cured nanocomposite, and the stress-intensity factor (KIC, up to 56.4%). The authors explained these values mainly by "the superior pulling-out effect of dispersed and tough MWCNT fiber upon the notched cracks". The σ_s, K_{IC}, and δ_C values can be balanced by incorporation of MWCNTS treated with acids. Additional short carbon fibers provoked a negative effect, meanwhile nanophase carbon black improved the fracture toughness.

There are controversial results for cement paste with admixed CNT of up to 500 μm in length, showing an increase or decrease in flexural or compressive strength. A small increase in fracture energy and tensile strength of a CNT-reinforced cement paste with 50 × 50 μm was reported [25]. CNT clustering was proved to be the crucial factor for an increase in fracture energy and for an improvement in tensile strength. Data from **Table 1** demonstrate that even CNTs shorter than 10 μm can be effective in providing added strength. In addition, the general transport properties (i.e., water sorptivity, water permeability, and gas permeability) of carbon-nanotube/cement composites were investigated [26]. Carboxyl-functionalized MWCNTs (outside diameter < 8 nm, inside diameter 2-5 nm, -COOH content 3.86 wt%, length 10-30 μm) were dispersed into cement mortar to fabricate the CNT-reinforced composites, by applying ultrasonic energy, in combination with the use of surfactants (sodium dodecylbenzene sulfonate, SDS, and NaDDBS). Several coefficients of cement mortar (water permeability coefficient, water sorptivity coefficient, and gas permeability coefficient) can be decreased by adding even small quantities of MWCNT-COOH (0.2%), thus increasing the durability of composites.

CNT length, μm	*w/c*	Compressive strength, MPa		Flexural strength, MPa	
		Plain	CNTs	Plain	CNTs
<10	P 0.30	38.3	61.8	2.3	2.3
<30	P 0.30	–	–	9.2	12.6
<10	M 0.40	28.9	54.3	6.0	8.23
<500	P 0.45	52.3	62.1	6.7	8.4
<100	P 0.50	–	–	5.5	7.2
<5	P –	49	56	16	8

Table 1. Strength gain when CNTs are added to cement (P = paste, M = mortar). Reproduced with permission from Ref. [25].

In cement composites, carbon fibers (CFs) can be applied alone or in combination with carbon nanotubes. The behavior of reinforced cement mortar composites round bars with MWCNTs (0.5 wt%; diameter 10-30 nm, length 1-2 μm, surface area 350 m^2/g) and CFs (2.25 wt%; length of fiber 5 mm, fiber thickness 0.3 mm), ultrasonically dispersed in the cement matrix, was elucidated [27]. Composite round bars were tested under direct tension, in order to evaluate their mechanical properties, such as ultimate load, deflection criteria, and stress-strain behavior. It was shown that the load carrying capacity of composite bars, under direct tension, is substantially higher than the plain controlled bar. Resulted plain cement bars with 2.25% carbon fiber reinforcement showed 38% increase in the ultimate load, whereas plain cement bars with 2.25% carbon fibers and 0.5% MWCNT showed 54% increase in the load carrying capacity. The presence of both carbon fibers and MWCNTs mainly contributed to the improvement of tensile load carrying capacity. Higher failure strain capacity of 44% was observed in bars with randomly distributed CF and MWCNTs as compared with plain bars.

Allotropes and hybrids of carbon can be obtained *in situ* on clinker particles as supports. Thus, the formation of a hybrid material (carbon clinker/nanofiber, **Figure 4**) was achieved by direct interaction of carbon nanofibers on clinker supporting particles (conditions: 550°C, fluidized bed reactor, gaseous C_2H_2, and CO_2) [28]. This interaction resulted in an excellent CNFs/clinker matrix dispersion and a strong final composite. At the 0.4% CNFs concentration, the enhancement of the mechanical properties of the mortar resulted in an increase of more than 2.5-fold in the compressive strength. In a similar manner, carbon nanotubes were grown *in situ* by chemical vapor deposition on a Portland cement clinker, in order to produce a nanocomposite material using clinker as support and catalyst [29]. Iron sources such as iron ore, steel mill scale, and red mud were used as additional transition metal catalysts to increase the carbon nanotube content. Thus, the clinker contains tetracalcium aluminoferrite (C4AF phase) that can be used as a catalyst for CNT growth. In a CVD process using ethylene as carbon source, and pure PC clinker both as a support and catalyst of CVD process, a CNT yield of 4.03%, in mass of clinker-CNT nanocomposite, was obtained.

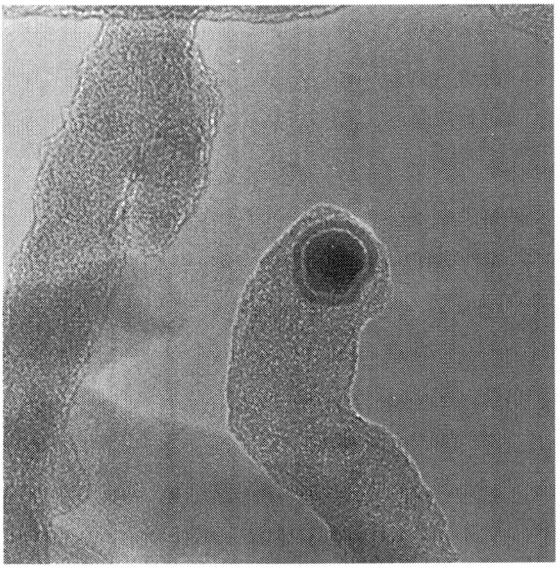

Figure 4. TEM image of the CNFs produced on the surface of clinker particles. Reproduced with permission.

4. Effects observed upon concrete reinforcement with carbon nanotubes

4.1. Effects of CNT functionalization, annealing, concentration, and water-cement ratio

–OH and –COOH functionalized CNTs have distinct physical properties and are more hydrophilic in comparison with pristine CNTs, so their effects on concrete properties can be different. The influence of functionalized MWCNT–OH and MWCNT–COOH on the impact resistance and compressive and flexural strengths of high–performance mortars (HPM) was studied [30]. The results of tests on reinforced high-performance mortar containing 0.1 wt% functionalized MWCNTs showed that the impact resistance, compressive strength, and flexural strength were 1400, 25.58, and 2% higher than those of HPM without MWCNTs. A more detailed comparison of distinct CNTs in cement matrix composites, prepared by addition of 0.5 wt% CNTs to plain cement paste, "was carried out [31]". Three different kinds of MWCNTs (**Table 2** and **Figure 5**) were used and compared: as-grown, annealed (high temperature (2100-2600°C); annealing effects on MWCNTs are described in [32, 33]), and carboxyl functionalized MWCNTs. The authors stated that high temperature annealing treatments remove lattice defects from the walls of CNTs, hence improving their mechanical

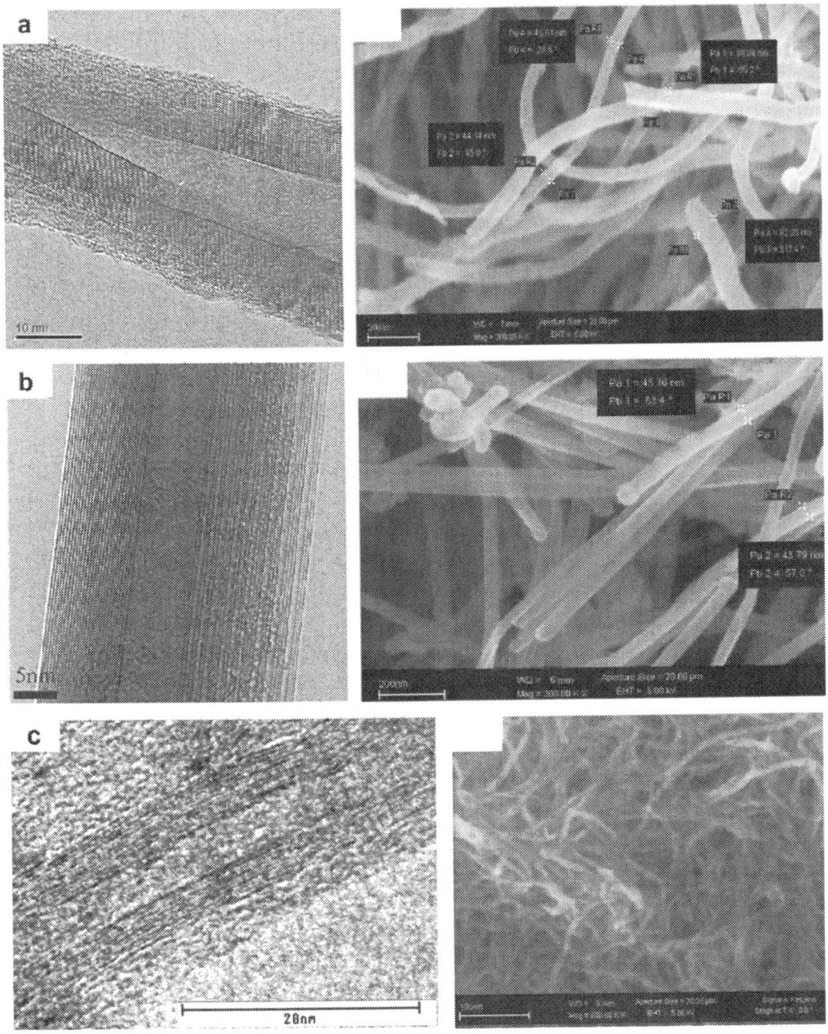

Figure 5. Electron microscopy analysis showing p-CNTs (a), a-CNTs (b), and f-CNTs (c). TEM images are reported on the left, while SEM images are on the right. Reproduced with permission.

strength. On the other hand, acid oxidative treatments increase chemical reactivity of pristine material, consequently chemical bonds between the reinforcement and the cement matrix are supposed to enhance the mechanical strength. Flexural and compressive tests showed a deterioration of the mechanical properties with functionalized MWCNTs, while a significant improvement is observed with both as-grown and annealed MWCNTs.

Property	p-CNTs	a-CNTs	f-CNTs
Deposition technique	CVD	CVD	CVD
Average diameter, nm	40–80	40–80	10–20
Average length, μm	400–1000	200–400	0.1–10
Carbon purity, wt%	>92	>99	>95
Metal oxide, impurity, wt%	<6	<1	<5
-COOH functionalization, wt%	0	0	<4

*p-CNTs: pristine MWCNTs; a-CNTs: annealed MWCNTs; f-CNTs: functionalized MWCNTs.

Table 2. Characteristics of the three different MWCNTs* dispersed in the cement. Reproduced with permission.

The piezoresistive property of the CNT/cement composite (**Figure 6**) shows an acidic treatment on CNTs (for their functionalization) and was used to explore its feasibility as an embedded stress sensor for civil structures, such as roadways, levees, and bridges [34]. It was shown that the electrical resistance of the CNT/cement composite changes with the compressive stress level, indicating the potential of using the CNT/cement composite as a stress sensor for civil structures (**Table 3**). It was also established that the dispersion-assistant surfactants could block the contacts among carbon nanotubes, thus impairing the piezoresistive response of the composite, while a higher CNT doping level could improve the sensitivity of the composite stress response. In addition, the effect of MWCNTs on strength characteristics and durability of concrete was investigated [35] by adding MWCNTs in different quantities (diameter 20-40 nm; length 1-10 μm; 0.015, 0.03, and 0.045%) with surfactants (super plasticizers-polycarboxylate 8H, 0.25% by weight of cement). Results showed an increase in compressive and splitting-tensile strengths of the samples with increasing MWCNT concen-

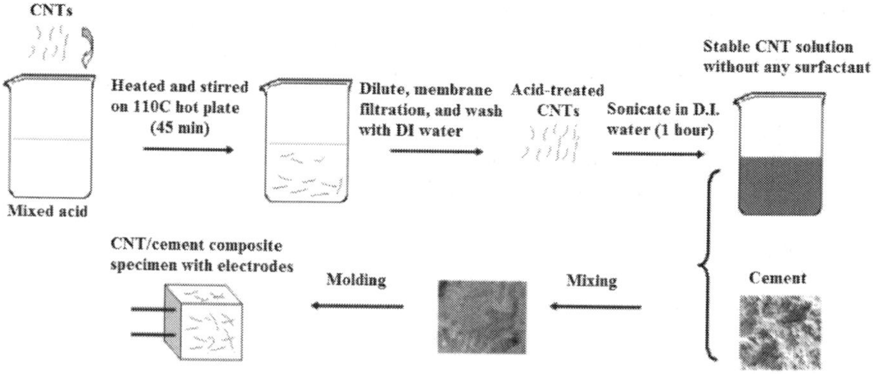

Figure 6. Illustration of the CNT/cement fabrication process based on the acid treatment of CNTs. Reproduced with permission.

tration (**Table 4**). By increasing the percentage of MWCNTs in the concrete, the water absorption is reduced to a greater extent, which helps in improving the concrete durability and water resistance.

CNT/cement composites	**Resistance change (5.2 MPa load)**	**Resistance change (8.6 MPa load)**
Cement only	0%	0%
0.06 wt% MWCNT	8.8%	10.3%
0.1 wt% MWCNT (by different methods)	5.0–9.4%	7.2–11.4%

Table 3. Comparison of electrical resistance changes of CNT/cement composites with different CNT doping levels and under different compressive loads. Reproduced with permission.

Concrete type	**Comp. failure load (kN)**	**Compressive strength (N/mm²)**	**% increase**	**Split tensile strength failure load (kN)**	**Split tensile strength**	**% increase**
Conventional concrete	875	38.22	–	160	2.27	–
0.015% MWCNTs	930	41.48	2.75	210	2.97	30.84
0.030 % MWCNTs	1010	45.18	16.38	235	3.30	45.37
0.045 % MWCNTs	1100	49.18	26.69	265	3.775	66.30

Table 4. Compressive and split tensile strength. Reproduced with permission.

Additionally, two mechanical properties (28-day compressive strength and flexural strength) of CNTs (diameter <1 nm up to 50 nm and length from 1 μm to 1 cm) and CNFs (diameter of 70-200 nm and length of 50-200 μm) cement composites were investigated [36]. Composites with 0.1 and 0.2% of CNTs and CNFs, and water/cement ratios between 0.35 and 0.5, were prepared under sonication. Both CNT and CNF composites demonstrated significant increase in compressive strengths, when compared to plain mortar control samples (maximum 154% for CNT and 217% for CNFs samples). Water/cement ratios in the range of 0.35-0.4 were found to produce the higher strengths, together with a 0.1% dosage rate for the CNTs/CNFs. It seems that the CNTs are better dispersed in the cement matrix than the CNFs, because a correlation between the flow test results and the compressive strengths was detected for the CNT samples. One other example that CNT concentration should not exceed certain levels was described for the heavy-weight oil well cements [37]. When only 1 wt% of CNTs was added to the cement slurry, the yield point and plastic viscosity increased by 8 and 5 times, respectively, while the free water and fluid loss of cement slurry were reduced by 85 and 70%, respectively. In addition, compressive strength of cement stone increased by 73.8%. However, the increase in the additive concentration leads to reduced compressive strength and Young's modulus, because of unsuitable dispersion of nanoparticles in the cement stone matrix; thus, an optimum level of CNTs should be used.

4.2. Length-to-diameter (l/d) aspect ratio effect

In addition to the effects of CNT length described above (**Table 1**), the intriguing effects of different concentrations of long (diameter <8 nm, length 10-30 μm) MWCNTs-high length/diameter aspect ratios of 1250–3750-and short MWCNTs (diameter 9.5 nm, length 1.5 μm)-aspect ratio of about 157-in cement paste were studied for 7, 14, and 28 days [38]. Both short and long MWCNTs were produced by the catalytic chemical vapor deposition (CCVD) process. **Figure 7** shows the relationships between the length-to-diameter aspect ratio and the surface-area-to-volume (SA/V) ratio for different lengths of SWCNTs, MWCNTs, carbon nanofibers, and carbon microfibers (CMFs). The authors noticed that only CNTs can provide a very high surface-area-to-volume (SA/V) ratio, which is one of the most important and desired elements in fiber-reinforced composite systems, in order to obtain the best and the most efficient materials. SWCNTs are the only materials that have a SA/V ratio that exceeds 2.0 nm^{-1}, especially when considering the ultra-long ones that have aspect ratios that can reach several millions. A higher SA/V ratio means a larger contact area between the fibers and the surrounding matrix, hence higher interaction with the matrix and more efficient reinforcing.

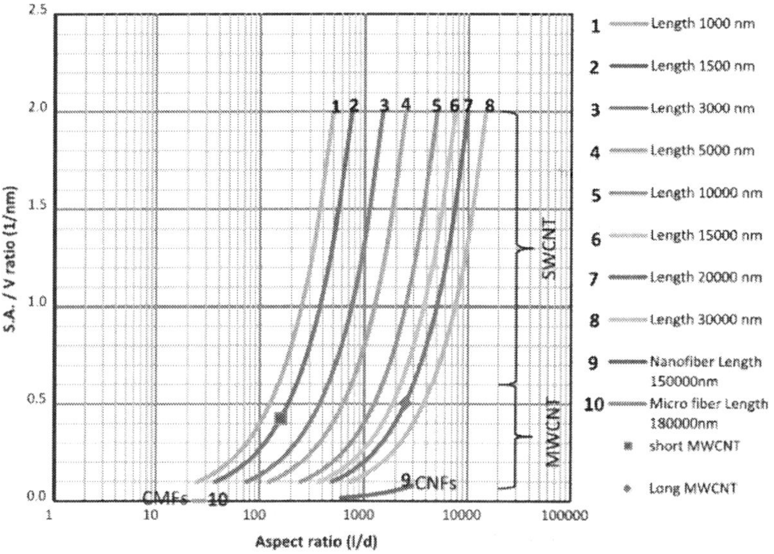

Figure 7. Length-to-diameter (l/d) aspect ratio effect on the surface-area-to-volume (SA/V) ratio for different lengths of SWCNTs, MWCNTs, CNFs, and CMFs. Reproduced with permission.

It was established that nanocomposites with low concentration of long MWCNTs yield comparable mechanical performance to the nanocomposites with higher concentration of short MWCNTs. **Figure 8** shows that the Young's modulus increases as the long CNTs' concentration decreases, whereas the Young's modulus increases as the short CNTs' concentration increases. Also, the authors noticed that low concentrations of long CNTs can lead to a much higher increase in the Young's modulus, as compared to higher concentrations of short CNTs. In addition, an example of the microcrack bridging by the MWCNTs, within the cement paste, is shown in **Figure 9**, clearly indicating that many CNTs are bridging the microcrack. Both

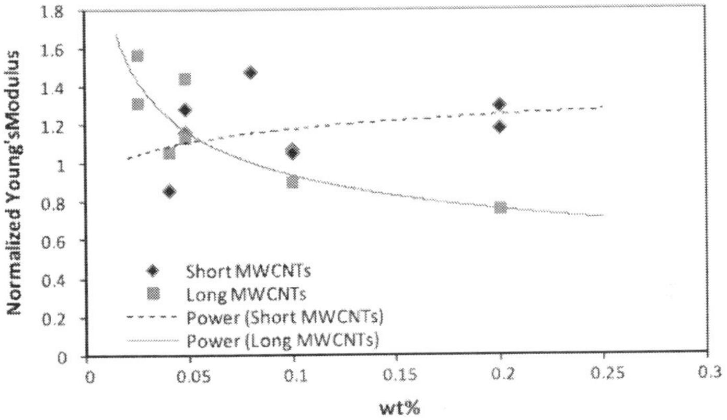

Figure 8. Variation of the normalized Young's modulus with the CNT concentration for two different aspect ratios; long and short. Reproduced with permission.

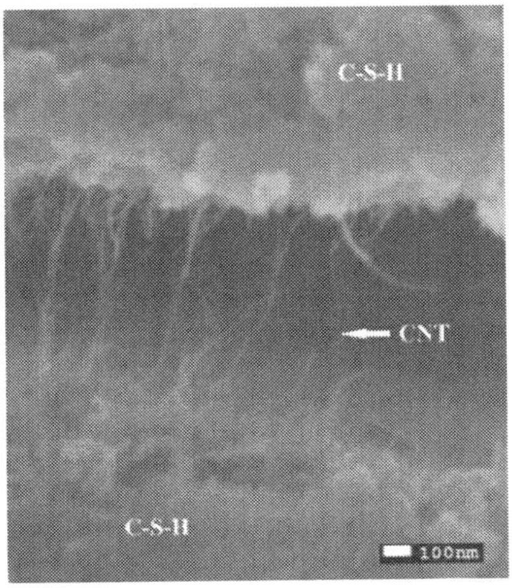

Figure 9. SEM image showing the microcrack bridging and breakage of the MWCNTs within the cement paste composite ((C–S–H) means calcium-silicate-hydrate). Reproduced with permission.

pull-out and breakage of CNTs can also be observed from images. In fact, the CNT breakage implies a good bonding between the CNT surfaces and the surrounding cement paste. However, it was noted by authors that the presence of CNTs affects the chemical reaction of the hydrated cement.

4.3. Effect of combination of CNTs with silica nanoparticles

Nanosilica (NS) is applied as an additive to improve the cement properties, giving a 15–20% increase of compressive strength [39]. Its combination with CNTs could lead to better results. Thus, the behavior of hardened cement paste reinforced with MWCNTs (0.75% by weight of cement; diameter 10–30 nm, length 1–2 mm) and NS (0.5%; particle size 10–20 nm) was investigated (**Figure 10**) after both MWCNTs and NS being dispersed using a ultrasonic energy method [40]. MWCNTs were used to reinforce the cement paste at the nanoscale, while NS was used to increase strength by forming an additional CSH gel. It was shown that the addition of MWCNTs and NS almost doubled the flexural strength in comparison with PC (the control beam).

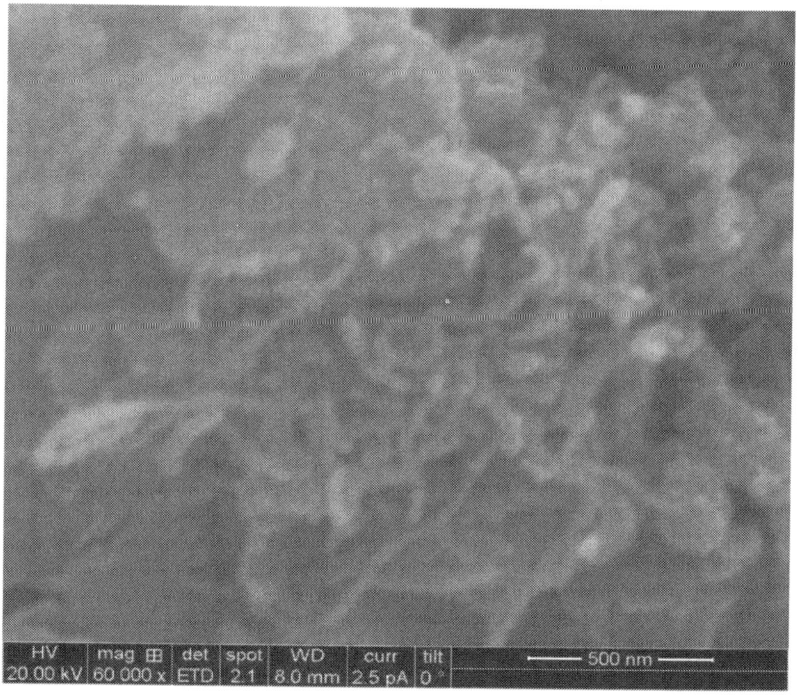

Figure 10. SEM image showing hardened cement paste modified with nanosilica and MWCNTs. Reproduced with permission.

5. Other specific studies of CNT/cement composites

5.1. Computational studies

Nanoscale and macroscale models were used to investigate the interfacial bond behavior of an individual CNT embedded in a cement matrix and the flexural response of a CNT/cement composite beam, respectively; also finite element (FE) models were applied to study the improvement in tensile strength and ductility of CNT-reinforced cement [41]. Industrial grade MWCNTs with a purity of 90 wt%, and a concentration of 0.25% by the total weight of cement paste, were ultrasonically dispersed in multiple steps before being added to cement (**Figure 11**). The FE models were successful in capturing the degradation in the CNT/cement bond strength and its effect on the flexural strength, ductility, and toughness of the composite. A comparison between the experimental and numerical behavior of the composite beams suggests an effective average shear strength value of 6.5 MPa at the interface of CNTs and cement matrix. Authors noted that, with a shear strength of up to 20.0 MPa, flexural strength, ductility, and toughness all increased with the interfacial bonding between CNTs and cement. This implies the significance of surface treatment of CNTs, which is commonly used to strengthen the bonding between CNTs and their surrounding matrix. As predicted by the FE analysis, increasing the shear strength from 6.5 to 20.0 MPa would increase the composite's flexural strength, ductility, and toughness by 141, 259, and 1976%, respectively. These results were significant, considering that the enhancement was provided by fibers alone, and highlight the great potential of surface-treated CNTs in reinforcing cement.

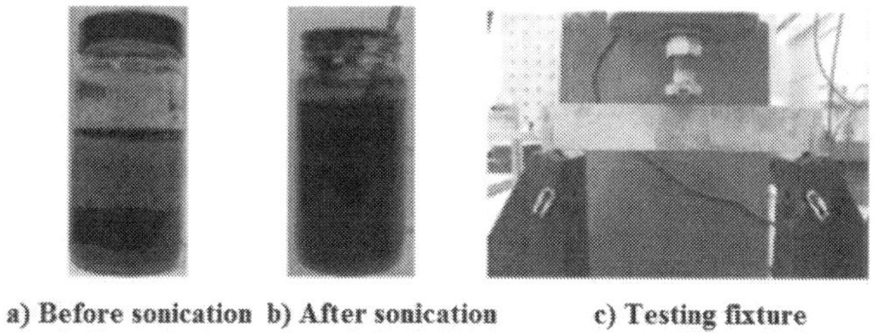

Figure 11. Experimental preparation for three-point bending tests of composite beams. Reproduced with permission.

5.2. TEM approach

A colloidal technique for transmission electron microscopy (TEM) of graphitic nanoreinforced cementitious composites was developed [42]. SWCNTs and MWCNTs (90 and 95% by mass with bulk densities of 0.14 and 0.27 g/cm^3 at 20°C, respectively; the inner diameters of the tubes were specified as 0.8–1.6 and 2–5 nm with average tube lengths of 5–30 and 10–30 μm for SWNTs and MWNTs, respectively) were functionalized using an acid etching technique to obtain stable aqueous suspensions to incorporate in the mix design of a cement paste. The functionalized nanoreinforcement and binding characteristics were observed at the nanoscale using high-resolution TEM imaging. Functionalized CNTs were found to be well distributed and preferentially associated with the cementitious matrix.

5.3. Study of electrical conductivity

A model for the electrical conductivity of heterogeneous systems based on cement and CNTs was developed [43]. The electrical properties of heterogeneous variances were shown to be dependent of such key parameters, such as the degree of aggregation of conductive particles and the electrical conductivity of a single unit. When the threshold concentration of electrical conductivity of CNT and graphite dispersions equals to 0.15, the electrical conductivity of CNT's systems were found to be 5–6 times higher than graphite. The authors emphasized that the CNTs have anomalous properties in comparison with graphite electrical dependencies due to the size and shape of the particles. It was also suggested that the small size effect of CNTs on the primary aggregation and the elongated shape of the particles affects the concentration at which a significant increase in electrical conductivity takes place.

5.4. Ultrasound tests

Tests with three different ratios (0.2, 0.4, and 0.6%, relatively high in comparison with those in reports described above) of carbon nanotubes (OD 5–60 nm, a length estimated from 5 to 30 μm) in concrete were carried out with the ultrasound technique. The measurement of the propagation delay of the signal through the specimens of cementitious materials with and without CNTs to reach the dynamic modulus of elasticity was carried out [44]. It was found that the mixture of 0.4% of CNTs was the mortar that showed the best performance, in comparison to the reference sample, reaching an increase of about 25% in the dynamic modulus of elasticity and a higher rate of structural compaction. The composites with addition of 0.2 and 0.6% of CNT did not show significant results for the same characteristics.

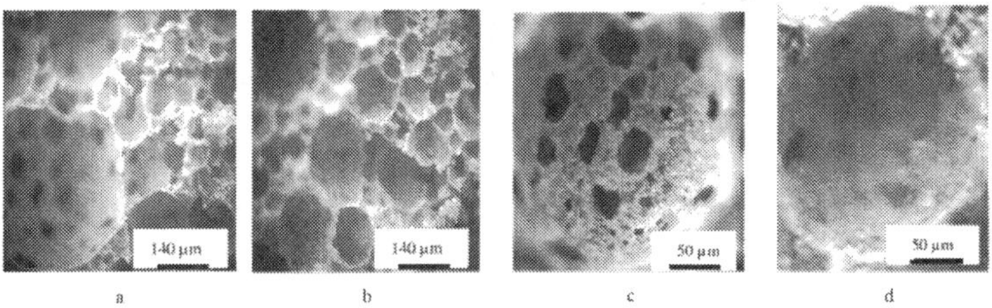

Figure 12. Cement foam concrete structure: (a) without nanotubes, (b) with 0.05% nanotubes (pore walls), (c) without nanotubes (perforated), and (d) stabilized with addition of 0.05% of nanotubes. Reproduced with permission.

5.5. Study of special foam-like cement

CNTs (80–90% of carbon, with a diameter of up to 100 nm and a length of up to 20 μm), agglomerated to fiber-shaped agglomerates with a diameter of up to 30 μm and a length of up to 10 mm, were synthesized using the method of stimulation of dehydropolycondensation and carbonization of aromatic hydrocarbons in chemical active environment (melts of aluminum, copper, nickel and iron salts), and used as a high strength dispersed reinforcement for the synthesis of nonautoclave concrete foam produced on the base of Portland cement (**Figure 12**) [45]. It was established that the use of CNTs (0.05% by mass), in the production of these concretes, allows to decrease its heat conductivity to 12–20% and increase its compressive strength to 70%.

6. Health impact

Considerable experimental data related to CNT toxicity at the molecular, cellular, and whole animal levels have been published [46]. The literature indicates considerable variability and uncertainty regarding the health impacts, reactivity, ecological effects, and environmental destination and transport of CNTs. In this respect, their addition to form high-performance cements should be also evaluated from the point of view of toxicity of resulted concrete, i.e., how the addition of CNTs may affect the environmental profile of cement. A recent report [47] evaluated hypothetical high-performance cements based on CNT reinforcement with a life cycle assessment (LCA) in order to compare the environmental impact of these new developments to traditional cements. It was established that the inclusion of CNTs increases considerably the environmental impact of cement production (**Table 5**). In addition, progress in research on these kinds of systems is largely hampered by the intrinsically hydrophobic nature of CNTs and their chemical incompatibility with cement hydrates. The authors proposed alternatives to CNTs as reinforcement for cements such as inorganic nanotubes or plastic nanofibers.

Impact category	Unit	1 kg CEM I 52.5 N	1 kg reinforced cement
Abiotic depletion of resources	kg Sb eq	2.184E-07	8.642E-06
Abiotic depletion of fossil fuels	MJ	1.485	56.719
Global warming, GWP	kg CO_2 eq	0.749	4.528
Ozone layer depletion, ODP	kg CRC-11 eq		6.396E-07
Human toxicity	kg 1,4-DB eq	0.067	1.353
Fresh water aquatic ecotoxicity	kg 1,4-DB eq	0.036	3.384
Marine aquatic ecotoxicity	kg 1,4-DB eq	100.751	5985.378
Terrestrial ecotoxicity	kg 1,4-DB eq	0.00082736	0.18839807
Photochemical oxidation	kg C_2H_4 eq	5.5717E-05	0.00125085
Acidification	kg SO_2 eq	0.00100558	0.03213931
Eutrophication	kg PO_4^{3-} eq	0.00022225	0.00745367

Table 5. Comparative impacts of standard and reinforced cements. Reproduced with permission.

7. Conclusions

The addition of CNTs and CNFs into cement forming nanocomposites leads to a considerable improvement of mechanical characteristics of concrete, since CNTs can act as effective bridges to minimize and limit the propagation of microcracks through the matrix. However, in order to reach the best effects, CNTs need to be well dispersed within the matrix to provide good bonding between them and the surrounding hydrated cement matrix, this being the main factor controlling concrete microstructure of cement composites. Addition of even 0.05% CNTs results in lower density, increased compressive strength, lower thermal conductivity, lower average pore diameter, and more homogeneous pore wall structure. The central idea of each cement-CNT experiment is to determine the best way to disperse CNTs. This can be reached by ultrasonication, use of surfactants (generally sodium dodecylbenzenesulfonate (SDBS), sodium deoxycholate (NaDC), Triton X-100 (TX10), GA, and cetyl trimethyl ammonium bromide (CTAB), whose dispersion ability varies considerably depending on molecular structure), use of cement admixtures (polycarboxylate, alkylbenzene, sulfonic acid, among many others), and covalent functionalization (commonly acid treatment to form -COOH and -OH functionalized CNTs). Combination of various chemical methods, such as the combination of surface functionalization with polymers, and other techniques, like an innovative method of fabricating cementitious nanocomposites through growth of CNTs onto the cement particles, also leads to good results. It has been finally established that the homogeneous CNT dispersion can be reached in the best form by dispersing them first in water, followed by mixing of the aqueous dispersion with mortar paste.

A series of effects have been discovered upon dispersion of carbon allotropes in concrete, showing the importance of diameter, length, and length-to-diameter ratio of CNTs, their concentration, functionalization, annealing, combination of CNTs with other nanomaterials, and water-cement ratio. We note that studies on CNT/cement nanocomposites are currently a relatively hot topic in nanotechnology [48–51] leading to best combinations of well-known materials having improved properties.

Author details

Oxana V. Kharissova, Leticia M. Torres Martínez and Boris I. Kharisov*

*Address all correspondence to: bkhariss@hotmail.com

1 Department of Physico-Mathematical Sciences, Autonomous University of Nuevo León, Monterrey, Mexico

2 Department of Civil Engineering, Autonomous University of Nuevo León, Monterrey, Mexico

3 Department of Chemical Sciences, Autonomous University of Nuevo León, Monterrey, Mexico

References

[1] Ferro, G.; Tulliani, J.-M.; Musso, S. Carbon nanotubes cement composites. *Frattura ed Integrità Strutturale*, 2011, *18*, 34–44.

[2] Sobolev K. Development of nano-SiO_2 based admixtures for high-performance cement-based materials. Progress report, CONACYT, Mexico, 2006.

[3] Kharisov, B.I.; Kharissova, O.V.; Leija Gitierrez, H.; Ortiz Méndez, U. Recent advances on the soluble carbon nanotubes. *Industrial & Engineering Chemistry Research*, 2009, 48(2), 572–590.

[4] Kharissova, O.V.; Kharisov, B.I.; de Casas Ortiz, E.G. Dispersion of carbon nanotubes in water and non-aqueous solvents. *RSC Advances*, 2013, 3, 24812–24852.

[5] Parveen, S.; Rana, S.; Fangueiro, R. A review on nanomaterial dispersion, microstructure, and mechanical properties of carbon nanotube and nanofiber reinforced cementitious composites. *Journal of Nanomaterials*, 2013, Article ID 710175, 19 pp.

[6] Olar, R. Nanomaterials and nanotechnologies for civil engineering. *Buletinul Institutului Politehnic din Iaşi*, 2011, *LIV*(LVIII), 109–117.

[7] Raki, L.; Beaudoin, J.; Alizadeh, R.; Makar, J.; Sato, T. Cement and concrete nanoscience and nanotechnology. *Materials*, 2010, 3, 918–942.

[8] Abinayaa, U.; Chetha, D.; Chathuska, S.; Praneeth, N.; Vimantha, R.; Wijesundara, K.K. Improving the properties of concrete using carbon nanotubes. *SAITM Research Symposium on Engineering Advancements*, 2014, (SAITM - RSEA 2014), 201–204.

[9] Li, G.Y.; Wang, P.M.; Zhao, X. Mechanical behavior and microstructure of cement composites incorporating surface-treated multi-walled carbon nanotubes. *Carbon*, 2005, *43*(6), 1239–1245.

[10] Cwirzen, A.; Habermehl-Cwirzen, K.; Penttala, V. Surface decoration of carbon nanotubes and mechanical properties of cement/carbon nanotube composites. *Advances in Cement Research*, 2008, *20*(2), 65–73.

[11] Luo, J.; Duan, Z.; Li, H. The influence of surfactants on the processing of multi-walled carbon nanotubes in reinforced cement matrix composites. *Physica Status Solidi A*, 2009, *206*(12), 2783–2790.

[12] Nasibulin, A.G.; Shandakov, S.D.; Nasibulina, L.I., et al. A novel cement-based hybrid material. *New Journal of Physics*, 2009, 11, Article ID 023013.

[13] Morsy, M. S.; Alsayed, S.H.; Aqel, M. Hybrid effect of carbon nanotube and nano-clay on physico-mechanical properties of cement mortar. *Construction and Building Materials*, 2011, *25*(1), 145–149.

[14] Nochaiya, T.; Chaipanich, A. Behavior of multi-walled carbon nanotubes on the porosity and microstructure of cement-based materials. *Applied Surface Science*, 2011, *257*(6), 1941–1945.

[15] Fraga, J.L.; del Campo, J.M.; García, J.A. Carbon nanotube - cement composites in the cement composites in the construction industry: 1952–2014. A state of the art review. In 2nd International Conference on Emerging Trends in Engineering and Technology (ICETET'2014), May 30–31, 2014, London, UK.

[16] Petrunin, S.; Vaganov, V.; Sobolev, K. The effect of functionalized carbon nanotubes on the performance of cement composites. NANOCON, October 16–18, 2013, Brno, Czech Republic, EU.

[17] Han, B.; Yu, X. Effect of surfactants on pressure-sensitivity of CNT filled cement mortar composites. *Frontiers in Materials*, 2014, 1, art. 27, 1–5.

[18] Wang, B.; Han, Y.; Pan, B.; Zhang, T. Mechanical and morphological properties of highly dispersed carbon nanotubes reinforced cement based materials. *Journal of Wuhan University of Technology - Materials Science Editorial*, 2013, 82–87.

[19] Kołtuńczy, E.; Nowicka, G. Effect of poly(sodium-4-styrenesulphonate) additives on properties of cement suspensions. In *Proceedings of International Scientific Conference "Surfactants and Dispersed Systems in Theory and Practice"*, Ed., K.A. Wilk, Palma Press, Wrocław, 2007.

[20] Li, C.; Zhu, H.; Wu, M.; Yan, M. Effect of dispersity of multi-walled carbon nanotubes on compression strength of cement. In *Fourth International Conference on Digital Manufacturing & Automation*, 2013.

[21] Yazdanbakhsh, A.; Grasley, Z.; Tyson, B.; Abu Al-Rub, R.K. Distribution of carbon nanofibers and nanotubes in cementitious composites. *Journal of the Transportation Research Board*, No. 2142, Transportation Research Board of the National Academies, Washington, DC, 2010, 89–95.

[22] Şener, S.; Çağlar, Y.; Belgin, Ç.M.; Şener, K.C. Modified Arcan tests for concrete with multi-walled carbon nanotubes. *Advanced Materials Letters*, 2014, 5(8), 429–434.

[23] Duan, Z.; Luo, J. Effect of multi-walled carbon nanotubes on the vibration-reduction behavior of cement. In *Proceedings of SPIE, International Conference on Smart Materials and Nanotechnology in Engineering*, Ed., S. Du, J. Leng, A.K. Asundi, 2007, *6423*, 64230R.

[24] Luo, J.L.; Duan, Z.; Xian, G.; Li, Q.; Zhao, T. Fabrication and fracture toughness properties of carbon nanotube-reinforced cement composite. *The European Physical Journal Applied Physics*, 2011, 53, 30402.

[25] Šmilauer, V.; Hlavácek, P.; Padevet, P. Micromechanical analysis of cement paste with carbon nanotubes. *Acta Polytechnica*, 2012, 52(6), 22–28.

[26] Han, B.; Yang, Z.; Shi, X.; Yu, X. Transport properties of carbon-nanotube/cement composites. *Journal of Materials Engineering and Performance*, 2013, *22*(1), 184–189.

[27] Hunashyal, A.M.; Tippa, S.V.; Quadri, S.S.; Banapurmath, N.R. Experimental investigation on effect of carbon nanotubes and carbon fibres on the behavior of plain cement mortar composite round bars under direct tension. *International Scholarly Research Network, ISRN Nanotechnology*, 2011, Article ID 856849, 6 pp.

[28] Nasibulina, L.I.; Anoshkin, I.V.; Semencha, A.V.; Tolochko, O.V.; Malm, J.E.M.; Karppinen, M.J.; Nasibulin, A.G.; Kauppinen, E.I. Carbon nanofiber/clinker hybrid material as a highly efficient modificator of mortar mechanical properties. *Materials Physics and Mechanics*, 2012, 13, 77–84.

[29] Ludvig, P.; Ladeira, L.O.; Calixto, J.M.; Gaspar, I.C.P.; Melo, V.S. In-situ synthesis of multiwall carbon nanotubes on Portland cement clinker. In *11th International Conference on Advanced Materials*, Rio de Janeiro, Brazil, September 20–25, 2009.

[30] Sahranavard, H.-K.; Abbasi, S.; Saeed, S.; Hasan, H.-K. Effect of multi-walled carbon nanotubes on mechanical properties of high-performance mortar. *Magazine of Concrete Research*, 2014, *66*(18), 948–954.

[31] Musso, S.; Tulliani, J.-M.; Ferro, G.; Tagliaferro, A. Influence of carbon nanotubes structure on the mechanical behavior of cement composites. *Composites Science and Technology*, 2009, *69*, 1985–1990.

[32] Ray, S.C.; Pao, C.-W.; Tsai, H.-M.; Chen, H.-C.; Chen, Y.-S.; Wu, S.-L.; Ling, D.-C.; Lin, I.-N.; Pong, W.-F.; Gupta, S.; Giorcelli, M.; Bianco, S.; Musso, S.; Tagliaferro, A. High-temperature annealing effects on multiwalled carbon nanotubes: electronic structure, field emission and magnetic behaviors. *Journal of Nanoscience and Nanotechnology*, 2009, *9*, 1–7.

[33] Kim, Y.A.; Hayashi, T.; Endo, M.; Kaburagi, Y.; Tsukada, T.; Shan, J., et al. Synthesis and structural characterization of thin multi-walled carbon nanotubes with a partially facetted cross section by a floating reactant method. *Carbon*, 2005, *43*, 2243–2250.

[34] Yu, X.; Kwon, E. A carbon nanotube/cement composite with piezoresistive properties. *Smart Materials and Structures*, 2009, *18*, 055010, 5.

[35] Madhavi, T.C.; Pavithra, P.; Baban Singh, S.; Vamsi Raj, S.B.; Paul, S. Effect of multi-walled carbon nanotubes on mechanical properties of concrete. *International Journal of Scientific Research*, 2013, *2*(6), 166–168.

[36] Yazdani, N.; Mohanam, V. Carbon nano-tube and nano-fiber in cement mortar: effect of dosage rate and water-cement ratio. *International Journal of Material Science (IJMSCI)*, 2014, *4*(2). www.ij-ms.org.

[37] Ghajari, A.; Gholinezhad, J.; Soltanian, H.; Alireza Mortazavi, S. An improvement to physical properties of heavyweight Oil well cements using carbon nanotubes. *Journal of Petroleum Science and Technology*, 2014, *4*(2), 10–19.

[38] Abu Al-Rub, R.K.; Ashour, A.I.; Tyson, B.M. On the aspect ratio effect of multi-walled carbon nanotube reinforcements on the mechanical properties of cementitious nanocomposites. *Construction and Building Materials*, 2012, *35*, 647–655.

[39] Sobolev, K.; Flores, I.; Hermosillo, R.; Torres-Martínez, L.M. Nanomaterials and nanotechnology for high-performance cement composites. In Proceedings of ACI Session on *"Nanotechnology of Concrete: Recent Developments and Future Perspectives"*, November 7, 2006, Denver, USA, 91–118.

[40] Hunashyal, A.; Banapurmath, N.; Jain, A.; Quadri, S.; Shettar, A. Experimental investigation on the effect of multiwalled carbon nanotubes and nano-SiO_2 addition on mechanical properties of hardened cement paste. *Advances in Materials*, 2014, *3*(5), 45–51.

[41] Chan, L.Y.; Andrawes, B. Finite element analysis of carbon nanotube/cement composite with degraded bond strength. *Computational Materials Science*, 2010, *47*, 994–1004.

[42] Aich, N.; Zohhadi, N.; Khan, I.A.; Matta, F.; Ziehl, P.; Saleh, N.B. Applied TEM approach for micro/nanostructural characterization of carbon nanotube reinforced cementitious composites. *Journal of Research Updates in Polymer Science*, 2012, *1*, 14–23.

[43] Lopanov, A.N.; Shukhov, V.G.; Fanina, E.A.; Kalchev, D.N. Electrical conductivity and aggregation of carbon nanotubes in a heterogeneous system based on cement. *Middle-East Journal of Scientific Research*, 2013, *17*(8), 1194–1199.

[44] Fernandes de Morais, J.; Naked Haddad, A. Analysis by ultrasound of the behavior of carbon nanotubes on cementitious composites. Journal of Nanotechnology and Advanced Materials, 2014, 2(2), 89–98.

[45] Yakovlev, G.; Keriene, J.; Gailius, A.; Girniene, I. Cement based foam concrete reinforced by carbon nanotubes. *Journal of Materials Science (Medžiagotyra)*, 2006, *12*(2), 147–151.

[46] Liu, Y.; Zhao, Y.; Sun, B.; Chen, C. Understanding the toxicity of carbon nanotubes. *Accounts of Chemical Research*, 2013, *46*(3), 702–713.

[47] Matanza, A.; Vargas, G.; Leon, I.; Pousse, M.; Salmon, N.; Marieta, C. Life cycle analysis of standard and high-performance cements based on carbon nanotubes composites for construction applications. In *World SB-4*, Barcelona, October 28–30, 2014.

[48] Manzur, T.; Yazdani, N. Strength enhancement of cement mortar with carbon nanotubes. Early results and potential. *Journal of the Transportation Research Board*, No. 2142, Transportation Research Board of the National Academies, Washington, DC, 2010, 102–108.

[49] Ghasemzadeh, H.; Akbari Jalalabad, E. Computing the compressive strength of carbon nanotube/cement composite. *International Journal of Civil Engineering*, 2011, *9*(3), 223–229.

[50] Sakulich, A.R.; Li, V.C. Nanoscale characterization of engineered cementitious composites (ECC). *Cement and Concrete Research*, 2011, 41(2), 169–175.

[51] Ubertini, F.; Luigi Materazzi, A.; D'Alessandro, A.; Laflamme, S. Natural frequencies identification of a reinforced concrete beam using carbon nanotube cement-based sensors. *Engineering Structures*, 2014, *60*, 265–275.

Chapter 8

Polycrystalline Diamond Thin Films for Advanced Applications

Irena Kratochvilova

Additional information is available at the end of the chapter

http://dx.doi.org/10.5772/64701

Abstract

The technological achievements in diamond thin film synthesis over the past decade subsequently led to the utilization of outstanding diamond properties and development of a wide range of applications in various fields of engineering. However, since most chemical vapour deposition (CVD) diamond films are polycrystalline, their characteristics strongly depend on their microstructure. As the number of possible applications for polycrystalline CVD diamond increases, there is constant development and enhancement of the film properties. Polycrystalline diamond in the form of thin films delivers further advantages over thicker polycrystalline layer, e.g., smoother surface, less deposition time and less light absorption. Furthermore, besides the relevant diamond properties, the suitability for applications also depends on various material parameters such as substrate nature, substrate dimensions, possibility of non-planar geometries, surface morphology, electrical conductivity, capability of device fabrication, electrochemical properties and cost.

Keywords: chemical vapour deposition polycrystalline diamond films, boron-doped polycrystalline diamond film, zirconium alloys protection against corrosion, water splitting

1. Introduction

Due to unique combination of mechanical, thermal and optical properties, manmade diamond is commonly used nowadays in a number of industrial applications. After decades of research and gaining general knowledge, new foreseen applications can be realized through improvements of diamond properties. Achievements in the preparation of synthetic diamond, in the form of thin films, allow for further research aimed at the development of modern applications.

A method, allowing preparation of polycrystalline diamond (PCD) films (**Figure 1**) under low-pressure and low-temperature regimes, was coined with the development of chemical vapour deposition (CVD). CVD apparatus is shown in **Figure 2**. A polycrystalline diamond film is formed during heteroepitaxial growth when the CVD deposition is carried out on a non-diamond substrate. The growth is achieved using small diamond crystal seeds dispersed onto a substrate. Due to eventual coalescence of different crystallites grown on variously oriented seeds (nucleation sites), a continuous PCD film with a columnar structure is formed. Such columnar structure is primarily made of diamond grains (crystallites) held together by thin layers of amorphous and graphitic inclusions located within the so-called grain boundaries. Overall, the PCD film thickness, the size of diamond grains at the surface and its roughness are determined at the moment when growth process is terminated [1].

Figure 1. Scanning electron microscopy top view images of the surface of the diamond film.

The technological achievements in the synthesis of diamond thin films over the past decade subsequently led to the utilization of outstanding diamond properties and development of a wide range of applications in various fields of engineering. However, since most CVD diamond films are polycrystalline, their characteristics strongly depend on their microstructure. As the number of possible applications for polycrystalline CVD diamond increases, there is constant development and enhancement of the films properties. Polycrystalline diamond in the form of thin films delivers further advantages over thicker polycrystalline layer, e.g., smoother surface, less deposition time and less light absorption. Furthermore, besides the relevant diamond properties, the suitability for applications also depends on various material parameters such as substrate nature, substrate dimensions, possibility of non-planar geometries, surface morphology, electrical conductivity, capability of device fabrication, electrochemical properties (given by the sp^3/sp^2 ratio) and cost. Due to benefits of CVD diamond films over single-crystalline form (low-cost, large area products), a variety of applications already available expanded beyond classical mechanical use of diamond. These include CVD polycrystalline diamond for optical, electronic and electrochemical applications, CVD diamond grades suitable for thermal management as well as PCD films for a number of further advanced applications [1–6].

In this chapter, new technological applications of polycrystalline diamond films are presented. First, we illustrate electrical applications of diamond films. The electrical properties of diamond films can be tuned by addition of dopants, resulting in p-type or n-type conductivity. Additionally, the electrical characteristics can be modified through surface termination with hydrogen, which gives rise to p-type surface conductivity. The origin of this effect is in strong covalent bonding between C and H that leads to the formation of two-dimensional surface layer for holes, spanning up to several nanometres below the diamond surface. Conducting microcrystalline, nanocrystalline and ultrananocrystalline diamond films features certain advantages and actively used as electrode material in electrochemistry. That includes chemical stability in aggressive media, optical transparency and wide working potential window. Therefore, diamond film electrodes nowadays find widespread use in various electroanalytical applications, e.g., chemical sensing, in vivo monitoring of bioactive spices, water purification and detection of organic pollutants.

Photoelectrochemical (PEC) cells for water splitting can offer a potentially inexpensive route for molecular hydrogen-fuel generation. In the frame of such application, nanocrystalline diamond has a potential to be utilized in PEC systems as a protective and electrically active coating for water oxidation. A unique diamond deposition technology allowing low temperature growth (ca. 450°C) was employed to fabricate boron-doped nanocrystalline diamond films on silicon solar cells. The goal of the study was to identify the properties of diamond films suitable for PEC application and to reveal the influence of diamond deposition process on the quality of silicon solar cells. Nanocrystalline boron-doped diamond (BDD) films possess grain boundaries containing a small fraction of non-diamond carbon impurities [4].

In the study of Ashcheulov et al. [4, 5], the conductivity of BDD polycrystalline films that were synthesized by microwave plasma-enhanced chemical vapour deposition on quartz substrates was studied. Conductivity is a complex physical quantity and in the case of BDD films there are many parameters that influence charge transport: (i) size of the diamond grains—the conductance of BDD films containing bigger grains is higher than the conductance of BDD films containing smaller diamond grains (for the same B content), (ii) diamond grain boundaries, (iii) setting of B defects and specific morphology of the layer changed by B addition and (iv) diamond grain surface layer can have specific setting and concentration of B defects. Boron-doped diamond is presently the subject of considerable interest as a promising electrode material [5]. This is a consequence of several technologically important properties that distinguish it from common electrode materials used in electrochemistry, such as gold or platinum. Polycrystalline BDD films possess physical properties similar to those of bulk diamond, which includes outstanding hardness, high hole mobility, high thermal conductivity and excellent resistance to radiation damage. Also, there is little evidence of degradation of BDD electrochemical activity with time. Due to these unique characteristics, to date, BDD electrodes are employed in a vast array of electrochemical applications. Additionally, then doped with high concentration of boron atoms, BDD films possess metallic conductivity and can also be optimized for light transparency [4, 5].

Nanocrystalline diamond layers were examined as anticorrosion protection of zirconium alloys nuclear fuel rods, operated in all commercial light-water and heavy-water nuclear reactors. Various corrosion tests were primarily focused on determining the ability of diamond layer to withstand emergency situations in nuclear reactors. The deposition of a polycrystalline diamond layer was carried out using linear antenna microwave plasma-enhanced chemical vapour deposition (LA-MW PECVD) apparatus (**Figure 2**) [1].

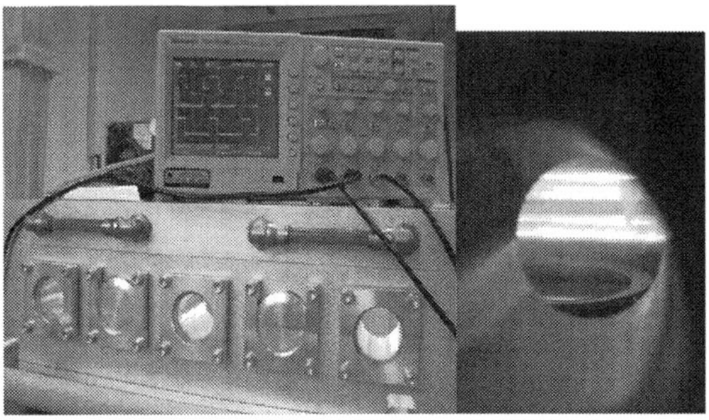

Figure 2. Linear antenna microwave plasma-enhanced chemical vapour deposition apparatus.

2. Doped nanocrystalline and epitaxial diamond layers for photovoltaic applications

It was showed that for hydrogen fuel generation based on water electrolysis by means of a light-driven photoelectrochemical process, boron-doped diamond is considered very promising material for electrodes [4]. The hydrogen produced using solar energy has been denominated as solar fuel. In essence, the photoelectrochemical process mimics photosynthesis by splitting water into H_2 and O_2. Nevertheless, water splitting cannot be achieved by sunlight directly because the water molecule cannot be electronically excited by sunlight photons. PEC water splitting is described by semiconductor's absorption of light photons with energies greater than its band-gap energy. Once photoexcited electrons and holes are generated in the conduction and valence bands, respectively, they separate and migrate towards the surface sites where they reduce and oxidize adsorbed water to produce gaseous oxygen and hydrogen. It has been demonstrated that various materials and composite structures can be utilized as photoelectrodes in PEC water-splitting processes. However, there are still obstacles to direct water splitting using PEC cells based on single absorber material: (i) lack of efficient light absorption due to wide band gap of semiconductors, (ii) efficiency reduction caused by overpotential and insufficient band alignments and (iii) corrosion of the semiconductor. Moreover, high cost of light-absorbing materials and co-catalysts (e.g., Pt, RuO_2 and IrO_2), which can split water with reasonable efficiencies prohibited their use on a large scale and over large areas required for light harvesting. Thereby, to raise the conversion efficiency, new low-cost and effective absorber materials and material compositions should be developed.

Silicon (Si) is an appealing material for photochemical processes due to a large number of photons that can be absorbed and converted. Further, Si is one of the most abundant elements in the Earth's crust. Thereto, Si is an attractive candidate for water-splitting platforms because of its prevalence in photovoltaic (PV) industries and predominance in electronics. Moreover, Si surpasses the majority of common semiconductor materials: the 1.1 eV band gap of Si allows effective light absorption from the UV to the near-infrared wavelengths region, thus resulting in a higher photocurrent and conversion efficiency.

Enabling Si as a photoelectrode for practical solar-fuel production from water mimics natural photosynthetic systems, where light absorption and catalytic reaction do not happen on the same material. Although, Si exhibits significant advantages over wide band-gap semiconductors (TiO_2, Fe_2O_3, WO_3, etc.) utilized in PEC devices, it has seldom been used in photocatalytic applications. The main reason is the significant corrosion of Si in non-acidic electrolytes; therefore, the realization of PEC device based on Si must account for its inherent corrosion. Recently, Si has been integrated as one of the main components of a novel concept of unwired macroscopic PEC device based on silicon solar cell of a standard structure (n-Si, p-Si and p^+-Si). In such system, p^+ side of Si solar cell is coated with corrosion resistant/conductive indium tin oxide (ITO) layer and catalyst for triggering the oxygen evolution reaction (OER).This approach enables fabrication of Si-based PEC devices with solar-to-fuel (STF) conversion efficiencies exceeding 10%. However, the scarcity and declining availability of indium forces to look for alternative corrosion resistant and electrically conductive material solutions.

Boron-doped diamond is presently the subject of considerable interest as a promising electrode material [4, 5]. This is a consequence of several technologically important properties that distinguish it from common electrode materials used in electrochemistry, such as gold or platinum. Polycrystalline BDD films possess physical properties similar to those of bulk diamond, which includes outstanding hardness, high hole mobility, high thermal conductivity and excellent resistance to radiation damage. Also, there is little evidence of degradation of BDD electrochemical activity with time. Due to these unique characteristics, to date BDD electrodes are employed in a vast array of electrochemical applications. Additionally, then doped with high concentration of boron atoms, BDD films possess metallic conductivity and can also be optimized for light transparency. It has been recently shown that a large potential window of BDD can be tuned by the fabrication of sp^2-abundant polycrystalline BDD films. Also, it has been reported, that BDD electrodes can be successfully utilized for disinfection of wastewater. On the whole, the above-mentioned properties make BDD electrodes an ideal candidate for implementation in PV-assisted electrochemical systems for solar-fuel generation by water decomposition and wastewater treatment.

Essentially, due to the favourable alignment of Si and BDD valence bands, a system based on Si solar cell and BDD electrode integrated in a PV-PEC device could be developed. Similarly to already reported PV-PEC systems with ITO layer, the rear side of the Si solar cell (p-type side) immersed into water has to be protected by electrically conductive and non-corrosive material, thus BDD layer acting as an anode has the capacity to render such demand.

Overall, this approach circumvents problems caused by the intrinsic properties of Si, such as corrosion in aqueous solution, minimizes the cost that is about one-third of the total PV system cost and could provide higher conversion efficiency as opposed to classical PEC systems.

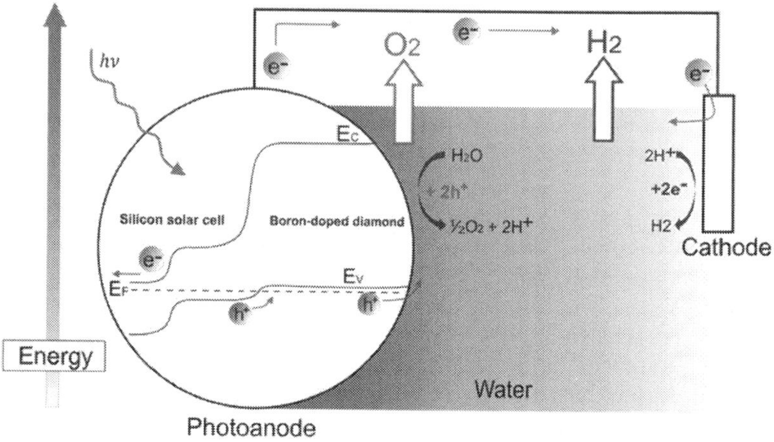

Figure 3. Schematic representation of the water-splitting system based on Si solar cell equipped with boron-doped diamond electrode deposited on the p-side of the Si solar cell with closed circuit on the externally wired cathode.

A schematic representation of the concept of the light-driven water-splitting process realized by PV-assisted PEC cell equipped with BDD electrode is illustrated in **Figure 3**. Here, the Si solar cell is responsible for the incident light absorption upon which photogenerated carriers are swept across the pn-junction. Due to a large band gap of highly boron-doped diamond films, set into contact with the p-side of Si, electrons are reflected and prohibited to move towards the BDD surface. Meanwhile, due to metallic doping of the BDD film, the Fermi level lies below the diamond valence band maximum that enables photogenerated hole carriers to pass freely through the BDD layer. Hence, holes are permitted to enter the water or electrolyte solution. Thus, the BDD film is responsible for water oxidation process and PV Si/BDD device could operate as photoanode, whereas the n-side of Si could be protected by a transparent glass layer may be further connected by a wire to a secondary electrode (cathode) that can be kept in a separate compartment to assist the separation of hydrogen and oxygen gases. Additionally, various OER and hydrogen evolution reaction catalysts can be deposited on the BDD surface and on the cathode, to facilitate the overall water-splitting process. Finally, due to the fact that commercially available single-junction Si solar cell is able to deliver potential of only ~0.7 V, which is not sufficient for triggering the water-splitting reaction (1.23 V), a tandem or a triple-junction of the c-Si solar cells connected in series would be necessary.

3. Conductivity of boron-doped polycrystalline diamond films: influence of specific boron defects

Diamond, as a carbon-based material, seems to be an exceptional choice for many applications, holding many promises due to its excellent properties, such as chemical inertness, wide electrochemical potential window and surface tunability by surface functionalization. Boron-doped diamond is a p-type semiconductor with E_g ~ 5.5 eV. Boron acceptor level in semiconducting diamond is about 0.36 eV from the valence band (EVB). Upon the metallic doping, The Fermi level (EF) shifts towards the valence band of diamond. Crystalline diamond has a strong and rigid isotropic structure due to its cubic crystal symmetry—sp^3-hybridized carbon atoms are bound by strong covalent bonds. Crystalline diamond is insulator with large band gap (more than 5 eV). Dopant atoms in the diamond matrices (usually boron or nitrogen) can strongly change electrical conductivity. The most widely used dopant is boron, conferring a p-type semiconducting character to diamond. The boron source can be a volatile compound introduced into the reactant gases, such as trimethyl borate or B2H6, or solid boron located near the substrate during the growth of the diamond film. The boron doping level is often in the range of 10^{18}–10^{20} atoms/cm^3 or a B/C ratio of about 10^5 to 10^3. Boron-doped diamond has randomly oriented microcrystallites with facets and grain boundaries characteristic of a polycrystalline material. B in a diamond lattice acts as an electron acceptor providing the diamond p-type semiconducting properties. The acceptor level introduced by B is quite deep 0.37 eV from the top of the valence band. At higher doping levels, conduction occurs by nearest neighbour and variable range hopping of holes between ionised B sites. Further addition of B results in a layer with conductivity comparable to metallic materials.

In the study of Ashcheulov et al. [5], the conductivity of boron-doped polycrystalline films was studied. BDD films were prepared by microwave plasma-enhanced chemical vapour deposition. Diamond is prepared by the decomposition of a mixture of methane (or other carbon-containing species) at pressures from 0.01 to 100 mbar and substrate temperatures of 250 to 800°C. If the chemical reaction takes place in an ionized plasma, the ionized gas requires much less energy (lower temperatures) to react. Ionization energy is generally provided by a high-frequency field.

Concerning BDD film conductivity, the following factors are important: (i) size of the diamond grains—BDD films containing bigger grains have bigger conductivity (for the same B content); (ii) grain boundaries with non-diamond carbon impurities; (iii) structures of boron defects and setting of the material; (iv) the surface of each diamond grain can have a higher concentration of B, allowing specific conduction. Resistance of boron-doped diamond layers as a function of boron concentration is shown in **Figure 4**.

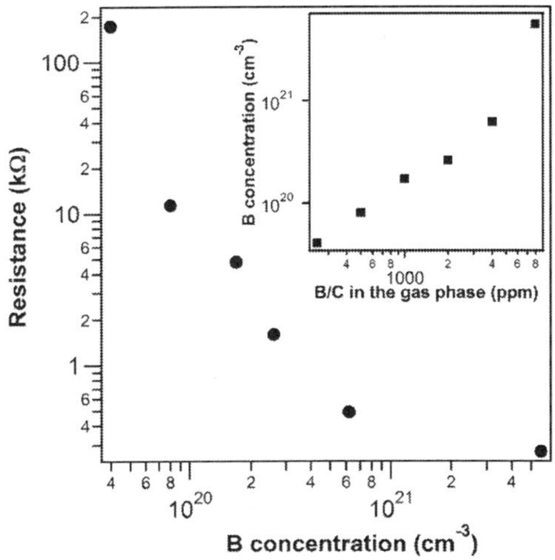

Figure 4. Resistance of boron-doped diamond layers as a function of boron concentration. Inset: Relation between the real boron concentration and the B/C ratio in the gas phase as measured by NDP.

Using a combination of theoretical and experimental techniques (plane-wave density functional theory, NDP, resistivity and Hall effect measurements, AFM, Raman spectroscopy), B defect location, vibration states, conductivity, free hole concentration inside the grains and hole mobility were investigated in [5]. Based on the analysis of specific structural, bond, interactions, charge localization parameters and spins, an explanation of the complex influence of specific B defects on the system charge transport mechanisms was set out:

- When the concentration of B in the PCD film is low, hopping conduction using the excited level of B defects is predominant. When the concentration of boron atoms in the diamond lattice is high enough so the wave functions of B overlap and form an impurity band passing to the top of the valence band metal-like conductivity behaviour of the material is reached.
- High boron concentration in diamond lattice can support complex defects (B dimers in deformed lattice). B dimers have high acceptor energy so electrons excitation from top of the valence band is very improbable. The impurity states do not merge with the valence band, and free carrier concentration is reduced. Moreover, due to compensated spins of 2s and 2p electrons in B dimmers, the probability of charge hopping is also reduced.

4. Protection of zirconium nuclear fuel cladding by polycrystalline diamond film of nanosize thickness

A 300 nm thick polycrystalline diamond layer has been used for the protection of zirconium alloy nuclear fuel cladding against undesirable oxidation with no loss of chemical stability and preservation of its functionality [2, 3, 6].

Reaction of fuel claddings with hot water steam triggered explosions of hydrogen gas in nuclear power reactors on several occasions, including the recent Fukushima accident. The probability of such accidents can be drastically lowered if the surface of the fuel claddings is coated with a suitable protective layer. In this chapter, we demonstrate that the 300 nm polycrystalline diamond layer on the surface of a Zr cladding rod improves the chemical stability of the cladding material while preserving its functionality.

Zirconium alloys are used in all commercial light-water (PWR, BWR and VVER) and heavy-water (CANDU) nuclear reactors. Nuclear fuel cladding (NFCs) are typically produced from zirconium alloys (with a common subgroup trademarked Zircaloy2), a material with very low absorption cross-section of thermal neutrons, high ductility, hardness and corrosion resistance [1–4]. Zircaloy2 serves faultlessly under standard operation conditions of the plant sealing the uranium oxide fuel pellets. Normal operating temperatures in the reactor is about 300°C. In the case of some accident at temperatures above 800°C the so-called high-temperature corrosion begins; in this case, the peeling of the oxide layer previously protecting the metal from oxidation occurs, which can result in mechanical failure of the system. It is strongly exothermic and high auto-catalytic reaction between zirconium and steam, during which the dissociation of steam molecules occurs and causes the formation of zirconium dioxide, hydrogen and releases a large amount of heat. The result of the reaction is not only the formation of hydrogen, which, as a combustible gas, is a serious risk in the case of a severe accident (like in recent Fukushima accident), but also the release of large quantities of heat, which further complicates the core cooling and strengthens the further course of the high-temperature oxidation of zirconium alloys. Finally, there is also the degradation of the fuel cladding, one of the protective barriers that may lead to its failure and subsequent leakage of highly radioactive fission products from nuclear fuel into the primary circuit. Many protective materials have been applied on the Zr alloys surfaces, but without any significant success. Just by applying a homogenous polycrystalline diamond layer on the surface of Zircaloy2, the Zircaloy2 cladding was successfully protected against undesirable changes and processes in the nuclear reactor environment. The layer of polycrystalline diamond will also prevent the reaction between zirconium and water steam. The deposition of polycrystalline diamond layer was carried out using LA-MW PECVD apparatus. In our recent work we demonstrated success: the 300 nm composite polycrystalline diamond layer with high crystalline diamond content and low roughness can protect zirconium alloy's nuclear fuel cladding against undesirable oxidation and consolidate its chemical stability while preserving its functionality. After ion beam irradiation (10 dpa, 3 MeV Fe^{2+}), the diamond film still showed satisfactory structural integrity with both sp^3 and sp^2 carbon phases— it was confirmed by Raman spectroscopy (**Figure 5**). The zirconium alloys under the carbon-based protective layer after hot steam oxidation test (1100°C, 30 min) differed from the original zirconium alloy's material composition only very slightly—the diamond coating increases the material resistance towards high-temperature oxidation by forming a protective composite structure. We also found that the presence of the PCD coating blocks hydrogen diffusion into the Zr surface and protects Zr material from degradation. By more closely monitoring the oxidation rate and phase change of PCD-coated Zircaloy2 fuel cell rods as a function of temperature, we found that even at temperatures above 850°C (Zr beta phase that is more opened for oxygen/hydrogen diffusion) the Zircaloy2 surface could be effectively protected against hot steam corrosion. PCD anticorrosion protection of Zr nuclear fuel rods can significantly (30%) prolong life of Zr material in nuclear reactors even above phase transition.

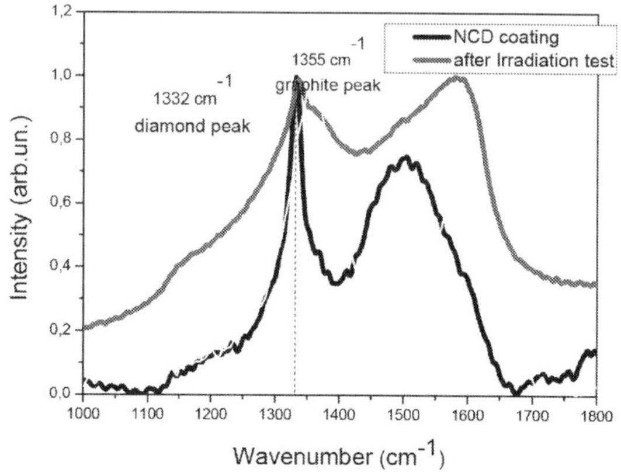

Figure 5. Comparison of Raman spectra of the PCD coating (black curve) and PCD after ion beam irradiation (red curve). Both sp^2 and sp^3 C (1332 cm^1 in Raman spectra) phases were present in the PCD layer.

After hot steam oxidation (950°C), a high level of structural integrity of the PCD layer was observed. Both sp^2 and sp^3 C (1332 cm^1 in Raman spectra) phases were present in the protective PCD layer. A greater relative weight gain of oxidized unprotected Zircaloy2 compared to that of the oxidized PCD-coated Zircalloy2 samples was supported by a thicker oxide layer of unprotected samples measured by impedance spectroscopy. The PCD layer blocks the hydrogen diffusion into the Zircaloy2 surface thus protecting the material from degradation. Hot steam oxidation tests confirmed that PCD-coated Zircaloy2 surfaces were effectively protected against corrosion. PCD film surface coating can prolong the service life of Zr nuclear fuel claddings in nuclear reactors.

Even under standard operating conditions in nuclear reactors (400°C hot water vapour), the polycrystalline diamond layer retains its original properties and participates both in the removal of heat released during the reactor operating mode, and further protects the coated surface against undesirable chemical reactions and changes in the composition of the structure related to the diffusion of hydrogen atoms from dissociated water molecules to zirconium alloy—as shown in **Figures 6** and **7**. The polycrystalline diamond layer further reduces especially adverse high-temperature chemical reactivity of the zirconium alloy surface, thereby also reducing high-temperature dissociation of water vapour molecules and subsequent formation of zirconium dioxide and explosive hydrogen. In the case of temperature-induced changes in zirconium alloy, the layer will benefit from the mixed nature of the protective carbon layer, which besides crystalline diamond grains with sp^3-hybridized carbon also contains flexible amorphous phase sp^2 from hybridized carbon capable of good adapta-

http://dx.doi.org/10.5772/64701

tion to volume changes/expansion of the metal substrate without compromising the protective layer integrity (**Figure 8**).

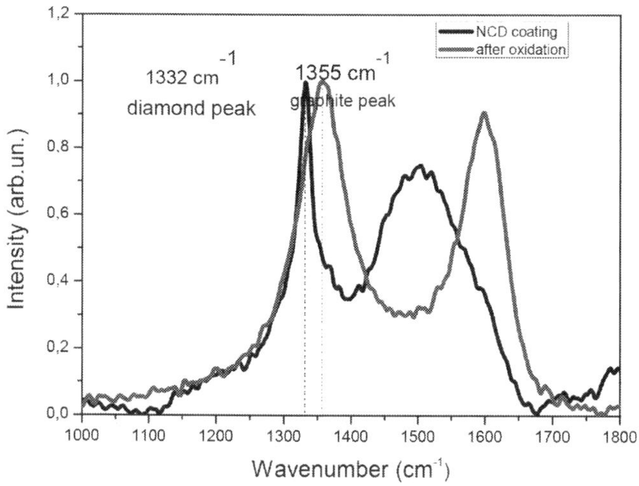

Figure 6. Comparison of Raman spectra of the PCD coating (black curve) and after hot steam oxidation (blue curve). Hot steam oxidation causes sp^2 transition of the layer.

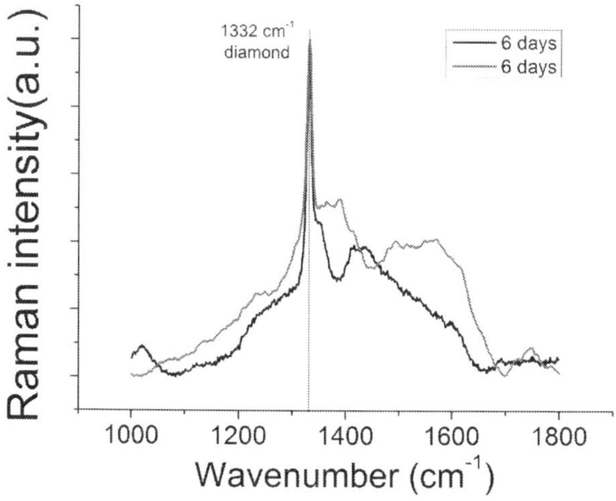

Figure 7. The 488 nm Raman spectra of PCD covered Zr alloy sample after 6 days in 400°C hot vapour/reactor testing. Spectra were taken from various parts of the tube. Both sp^2 and sp^3 C (1332 cm^1 in Raman spectra) phases were present in the PCD layer.

Figure 8. Left: A sample of PCD-coated zirconium fuel cladding; right: a lengthwise cut of a sample of PCD-coated zirconium fuel cladding, which after being exposed to water vapour at 400°C for 3 days was completely straightened. The PCD layer has withstood even this strain without compromising its integrity. The scratch in the right lower corner (right) was made using a diamond point pen.

5. Conclusions

From carbon nanomaterials especially polycrystalline diamond films have attracted more and more interest due to their unique electrical, optical and mechanical properties, which make them widely used for different applications. Diamond survives high temperatures, does not degrade over time and (important for the nuclear fuel cladding) has perfect neutron cross-section properties. Moreover, polycrystalline diamond layers consisting of crystalline (sp^3) and amorphous (sp^2) carbon phases could have suitable thermal expansion. Polycrystalline diamond films can be used in a wide range of applications: high boron- and phosphorus-doped nanocrystalline diamond layers that can have a significant impact in the industrial use of diamond in electrochemical applications. Boron-doped nanocrystalline diamond-coated silicon solar cells can be used for direct photoelectrochemical water splitting with high effectivity. In such a system, boron-doped diamond electrodes are responsible for water decomposition. Diamond could also be considered as a prospective material for electrodes and organic solar cells. Diamond layers were used as anticorrosion protection for zirconium alloys applied in nuclear reactors. Polycrystalline diamond anticorrosion protection of Zr alloys can significantly prolong life of Zr material in nuclear reactors even at temperatures above Zr phase transition temperatures. Zr alloy's oxidation and hydrogenation lead to limitation criteria in justifying fuel rod operability—rods containing even non-irradiated nuclear fuel must be fired due to Zr alloy surface destruction by oxidation/corrosion in reactor. Thus, the protection of zirconium fuel rods' surfaces against corrosion in nuclear reactors' environment can protect nuclear fuel to important extent.

http://dx.doi.org/10.5772/64701

Author details

Irena Kratochvilova

Address all correspondence to: krat@fzu.cz

Institute of Physics, Academy of Sciences Czech Republic v.v.i, Prague, Czech Republic

References

[1] F. Fendrych, A. Taylor, L. Peksa, I. Kratochvilova, J. Vlcek, V. Rezacova, V. Petrak, Z. Kluiber, L. Fekete, M. Liehrand and M. Nesladek: *Growth and characterization of nanodiamond layers prepared using the plasma-enhanced linear antennas microwave CVD system,* J. Phys. D: Appl. Phys. 43(2010), 374018.

[2] I. Kratochvílová, R. Škoda, J. Škarohlíd, P. Ashcheulov, A. Jäger, J. Racek, A. Taylor and L. Shao: *Nanosized polycrystalline diamond cladding for surface protection of zirconium nuclear fuel tubes,* J. Mater. Process. Technol. 214(2014), 2600–2605. IF 2.1

[3] P. Ashcheulov, R. Škoda, J. Škarohlíd, A. Taylor, L. Fekete, F. Fendrych, R. Vega, L. Shao, L. Kalvoda, S. Vratislav, V. Cháb, K. Horáková, K. Kůsová, L. Klimša, J. Kopeček, P. Sajdl, J. Macák, S. Johnson and I. Kratochvílová: *Thin polycrystalline diamond films protecting Zirconium alloys surfaces: from technology to layer analysis and application in nuclear facilities,* App. Surf. Sci. 359(2015), 621–628.

[4] P. Ashcheulov, M. Kusko, F. Fendrych, A. Poruba, A. Taylor, A. Jäger, L. Fekete, I. Kraus and I. Kratochvílová: *Nanocrystalline diamond on Si solar cells for direct photoelectrochemical water splitting,* Phys. Status Solidi A. 1–6(2014). DOI 10.1002/pssa.201431177

[5] P. Ashcheulov, J. Sebera, A. Kovalenko, V. Petrak, F. Fendrych, M. Nesladek, A. Taylor, Z. Vlckova Zivcova, O. Frank, L. Kavan, M. Dracinsky, P. Hubik, J. Vacik, I. Kraus and I. Kratochvilova, *Conductivity of boron-doped polycrystalline diamond films: influence of specific boron defects,* Eur. Phys. J. B. 86(2013), 443.

[6] P. Ashcheulov, R. Škoda, J. Škarohlíd, A. Taylor, F. Fendrych and I. Kratochvílová: *Layer protecting the surface of zirconium used in nuclear reactors,* Recent Patents Nanotechnol., 10(2016).

Section 3

Nanofluids and Case Studies on Carbon Nanostructures

Chapter 9

Nanofluids Based on Carbon Nanostructures

Hammad Younes, Amal Al Ghaferi, Irfan Saadat and
Haiping Hong

Additional information is available at the end of the chapter

http://dx.doi.org/10.5772/64553

Abstract

A nanofluid consists in a liquid suspension of nanometer-sized particles. These fluids may contain (or not) surface-active agents to aid in the suspension of the particles. Nanometer-sized particles have higher thermal conductivity than the base fluids. Oxides, metals, nitrides, and nonmetals, like carbon nanotubes, can be used as nanoparticles in nanofluids. Water, ethylene glycol, oils, and polymer solutions can be used as base fluids. In this chapter, we summarize the recent studies of using CNTs and graphene to improve the thermal conductivity of nanofluids. Moreover, we refer to the studies about the effect of using magnetic fields on enhancing the thermal conductivity of nanofluids. Too much discrepancy about thermal conductivity of nanofluids can be found in the literature. For carbon nanofluids, unfortunately, no significant improvements on thermal conductivity are observed using low concentrations. Different improvement percentages have been reported. This variation in the thermal conductivity can be attributed to many factors, such as particle size temperature, pH, or zeta potential. We believe that more research efforts need to be made in order to, first, improve the thermal conductivity of nanofluids and, second, assess the effect of the different parameters and conditions on the thermal conductivity of nanofluids.

Keywords: CNTs, graphene, nanofluids, thermal conductivity, alignment and magnetic field

1. Introduction

The enhancement of the efficiency of heat transfer fluids has been of great interest to scientists and engineers for the past two decades. Conventional heat transfer fluids, such as mineral oil, water, and ethylene glycol (EG), have poor heat transfer properties, compared to solids. In recent years, nanofluids have been investigated as potential heat transfer fluids [1–11].

The conventional coolants/fluids used in radiators and engines have poor heat transfer properties and contain millimeter- or micrometer-sized particles to improve the heat transfer properties that can clog new cooling technologies that include microchannels. The use of fluids with better heat transfer can allow engines to run at optimal temperatures and allow the use of smaller and lighter radiators, pumps, and other components. Smaller radiators allow a smaller front end of large vehicles, such as trucks. Fluids with better heat transfer can also lead to the production of lighter vehicles with better fuel economy, since less pumping power is needed. Although many factors such as pH, particle size, and zeta potential can affect the thermal conductivity (TC) of nanofluids, they are not promising to improve thermal conductivity values [12–16].

Wensel et al. [17] found that using low concentrations of the nanoparticles do not enhance the TC effectively. For example, using of 4–5 vol.% metal oxides increases the thermal conductivity by 10–20% only, which is not the significant increase that is needed.

2. Carbon-based nanomaterials

In this chapter, two carbon-based nanomaterial nanofluids will be discussed: carbon nanotube (CNT) and graphene-based nanofluids and their thermal conductivity.

2.1. Carbon nanotubes

Carbon nanotubes (CNTs) have attracted significant attention since they were discovered by Iijima in 1991 [18]. The sp^2 carbon–carbon bond in the plane of the graphene lattice is the strongest among of all chemical bonds. CNTs can be obtained with different numbers of walls, including single-walled CNTs, double-walled CNTs, and multiwall CNTs. They have very good thermal conductivity along the length of the tubes. It has been shown that the thermal conductivity of CNTs is unusually high (e.g., 3000 W/mK) for MWNTs [19] and even higher for SWCNTs (6000 W/mK). The thermal conductivity of CNTs is more than seven times higher than that of copper, which is well known for having a good thermal conductivity of 385 W/Mk [19–22]. Because of their remarkable thermal, electrical, mechanical properties, and low density, carbon nanotubes are ideal materials for reinforcement in composites [23–27], sensors [28], and many other applications, such as thermal managements [29–31].

2.1.1. Carbon nanotubes and carbon nanotubes metal oxide-based nanofluids

Due to the high thermal conductivity of CNTs, researchers started to use them to prepare nanofluids in order to improve the heat transfer. CNTs tend to aggregate into groups due to the large surface energy (strong van der Waals attractions between individual tubes) [32]. Therefore, a variety of approaches has been introduced to decrease the nanotubes agglomeration, namely, the modification of their chemistry through noncovalent adsorption using surfactants [33–36], covalent (functionalization) by chemical modification [37–42], and metal coating (like Ni-coated SWNTs). **Table 1** summarizes the work carried out so far dealing with the use of CNTs in nanofluids. Xie et al. [52] investigated the functionalized carbon nanotubes and found out that the thermal conductivity enhancement increased with the increase in nanotubes loading. In addition, it was found that the thermal conductivity decreased with the increase in thermal conductivity of the base fluid. Chen et al. [46] studied the effect of MWNTs on the thermal conductivity enhancement of ethylene glycol-based nanofluid and found that the enhancement reached up to 17.5% at 0.01 volume fraction. Ding et al. [49] observed a significant enhancement of the convective heat transfer in comparison with pure water as the working fluid, which depends on the flow condition, CNT concentration, and the pH level (although the latter is small). Grag et al. [44] studied the effect of ultrasonication on the thermal conductivity of MWCNT fluids and found that the thermal conductivity was enhanced by 20% when the ultrasonication was used for 40 min in 1 wt.% MWCNT concentration, in a 130 W, 20 kHz ultrasonication environment.

Reference	Material	Base fluid	Concentration	Characterization	TC improvement	Stabilizing agent	Stability
Phuoc et al. [43]	MWCNTs	Water	1.43 vol.%	Density and sound analyzer	13%	Chitosan	Stable for 45 days
Garg et al. [44]	MWCNTs	Water	1 wt.%	TEM,	20%	Gum Arabic	Stable
Meng et al. [45]	MWCNTs	EG	4 wt.%	TEM, UV-VIS	24.3%	Functionalization	Stable
Chen et al. [46]	MWCNTs	Water	1 vol.%	TEM,	17.5%	NA	Stable for many months
Chen and Xie [47]	SWCNTs	Water	0.2 vol.%	TEM	15.6%	Functionalization	Stable
	DWCNTs				14.2%		
	MWCNTs				12.1%		
Choi et al. [12]	MWCNTs	PAO	1.0 vol.%	TEM	160%	NA	Stable
Assael et al. [48]	MWCNTs	Water	0.6 vol.%	HR-TEM	38 %	Sodium dodecyl sulfate (SDS)	Stable
Ding et al. [49]	MWCNTs	Water	0.6 vol.%	TEM, SEM	79%	SDS, gum Arabic	Stable
Shaikh et al. [50]	CNT	PAO	1.0%	SEM	161%	Functionalization	Stable
	EXG	PAO	1.0%	SEM	131%		Stable
	HTT	PAO	1.0%	SEM	104%		Stable
Baby and Sundara [51]	Ag/ (MWCNT-HEG	EG	0.04 vol.%	SEM, XRD	8%	NA	Stable

PAO: poly (α-olefin) oil.

Table 1. Current research of carbon nanomaterials-based nanofluids.

Phuoc et al. [43] studied the thermal conductivity, viscosity, and stability of nanofluids made of MWCNT dispersed in water. The nanotubes have an outside diameter of 20–30 nm, inside diameter of 5–10 nm, length of 10–30 mm, and average density of 2.1 g/cm^3. Different weight ratio of Chitosan surfactant, namely, 0.1, 0.2, and 0.5 wt.%, was used to disperse MWCNTs in water. It was found that Chitosan dispersed and stabilized MWNTs in water efficiently; as only 0.1 wt.% Chitosan surfactant could disperse and stabilize 3 wt.% MWCNTs in water. The thermal conductivity of nanofluids containing 0.5–3 wt.% MWCNTs was enhanced from 2.3 to 13%. However, the enhancement of the thermal conductivity was independent of viscosity of the base fluid, which shows that the particle velocity does not have a significant effect on the thermal conductivity. **Figure 1** shows the dispersion behavior of the MWCNTs in deionized (DI) water (DW).

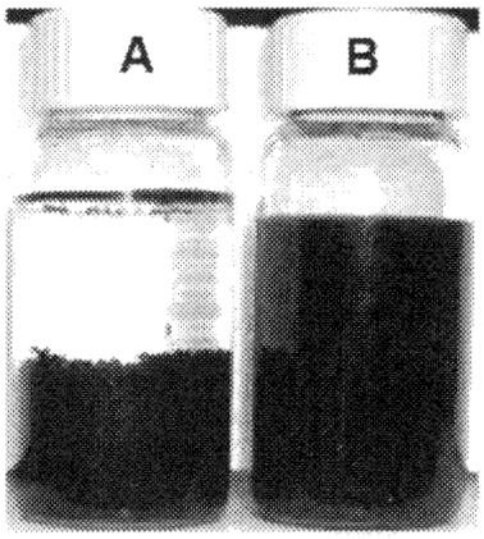

Figure 1. Nanofluid prepared by dispersing MWCNTs in deionized water (A) shows that MWNTs deposited at the bottom of the container and (B) shows that MWNTs maintain good dispersion: From Ref. [43].

Garg et al. [44] reported that 1 wt.% MWCNT increased the thermal conductivity up to 20% at 35°C. Gum arabic was used as a surfactant to stabilize the nanofluids. The study revealed that the thermal conductivity of CNT nanofluids increased considerably after 24°C. Meng et al. [45] found a 25.4% enhancement of the thermal conductivity of CNT glycol nanofluids with 4.0% mass fraction at room temperature. **Figure 2** shows that the transmittances from 200 to

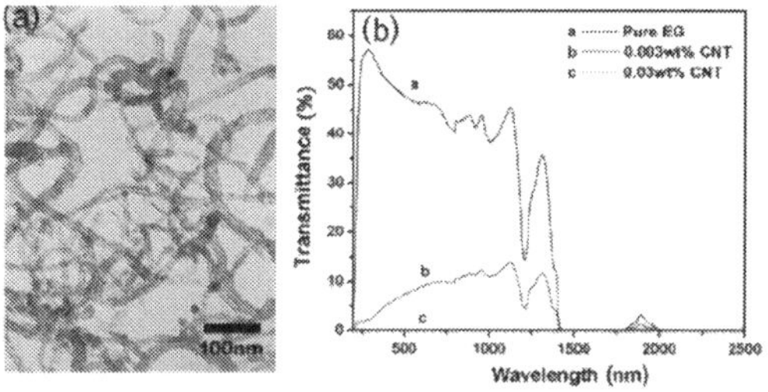

Figure 2. TEM image of HNO_3-treated CNTs (A) and UV–Vis–NIR transmittance spectra (B) of the CNTs glycol nanofluids. From ref. [45].

1500 nm, the values drop from 50% to about 10%, indicating that a small amount of CNTs results in significant absorption enhancement of light.

Chen et al. [46] reported that 1 vol.% of MWCNTs without surfactant, placed in ethylene glycol-based nanofluid, yielded an enhancement of 17.5%. A mechano-chemical reaction method was used to enhance the MWCNTs dispersibility for producing a CNT nanofluid. **Figure 3** shows the thermal conductivities of nanofluids as a function of nanotube loadings in two different fluids: deionized water and ethylene glycol. The enhancement using ethylene glycol as fluid and without surfactant at 1 vol.% of MWCNTs is higher than the enhancement using DW as fluid and at 1 vol.% of MWCNTs.

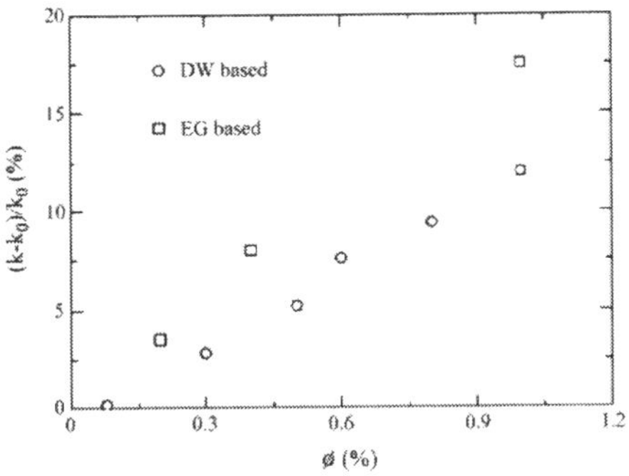

Figure 3. Thermal conductivities of nanofluids as a function of nanotube loadings. From ref. [47].

Chen and Xie [47] investigated nanofluids made of SWCNTs and DWCNTs in water-based fluids, respectively. Those CNTs were functionalized by a wet-mechano-chemical reaction method in order to obtain a stable dispersion. The study revealed that the thermal conductivity enhancement was 15.6, 14.2, and 12.1% for SWCNT, DWCNT, and MWCNT at a volume fraction of 0.2%, respectively.

Choi et al. [12] reported that the enhancement of thermal conductivity at a 1% volume fraction of MWCNTs in poly (α-olefin) (PAO) oil was 160%, compared to the base fluid at room temperature.

Assael et al. [48] found a 38% enhancement of the thermal conductivity in 0.6 vol.% MWNTs nanofluids in which water was a base fluid and SDS surfactant was used to stabilize the nanofluids. Ding et al. [49] confirmed that 1.0 vol.% of MWCNT dispersed in water, with SDBS surfactant, lead to an enhancement of 79% at 30°C. Shaikh et al. [50] prepared nanofluids by dispersing CNTs, exfoliated graphite (EXG), and heat-treated nanofibers (HTT) in PAO oil. The thermal conductivity for 1% CNTs, EXG, and HTT was enhanced by 161, 131, and 104 %, respectively. Baby and Sundara [51] reported on the synthesis of silver nanoparticle decorated MWCNT-graphene mixture and dispersed the (Ag/(MWNT-HEG)) composite in ethylene glycol without surfactant. They found that only 0.04 vol.% of the nanoparticles enhanced the thermal conductivity by ~8% at 25°C.

2.2. Graphene

Graphene, the first ever stable two-dimensional (2D) honeycomb lattice of sp^2-bonded carbon atoms, has attracted the interest of many researchers and is very promising for a wide range of applications due to its spectacular electronic, mechanical, and optical properties. Since few layers of graphene were isolated for the first time from graphite, using a tape in 2004 [53], the graphene production rate has increased rapidly in order to synthesize large-area and high-quality graphene films, compatible with various applications. Actually, the history of graphene dates back to the 1960s, when single-atom plane of graphite, termed as graphite layer, was isolated. For several decades, the existence of 2D atomic structures remained as an unaccepted concept. Now the term of graphene, the basic block of graphite, is widely used to describe the 2D monolayer of carbon atoms with a honeycomb structure. Graphene lattice consists of repetitive two identical carbon atoms, A and B, that together form one unit cell (**Figure 4**). The distance between each carbon atom and its neighbors is 1.42 Å [54]. **Figure 5** shows first Brillouin zone and band structure of graphene using tight binding approximation. *K* and *K'* are Dirac points, the transition between the valence and the conduction band.

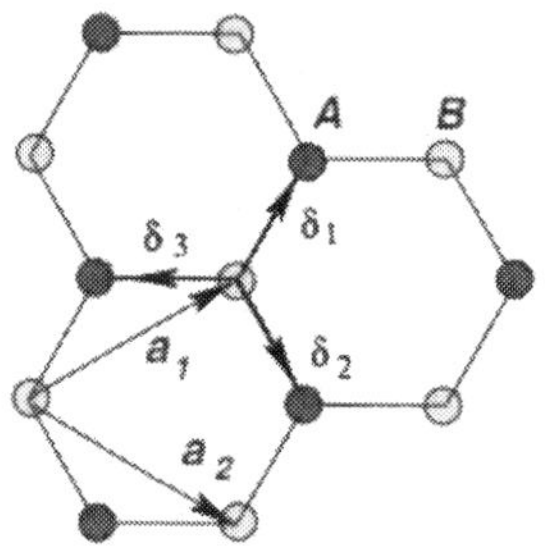

Figure 4. Graphene lattice.

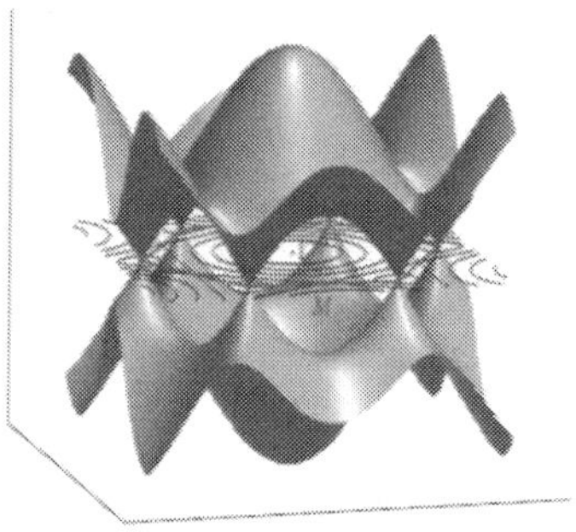

Figure 5. First Brillouin zone and band structure of graphene using tight binding approximation. *K* and *K'* are Dirac points, the transition between the valence and the conduction band. Reproduced from [54].

2.2.1. Graphene and graphene metal oxide-based nanofluids

The high thermal conductivity of graphene encouraged researchers to use it to prepare nanofluids with super high thermal conductivity. Yu et al. [55] made stable nanofluids by dispersing graphene oxide nanosheets in ethylene glycol and found out that the thermal conductivity of the nanofluids was higher than that of the ethylene-glycol base fluid. The thermal conductivity of the nanofluids was enhanced up to 61.0% at 5.0 vol.% of graphene nanosheets (GNs), and almost constant for 7 days, indicating that the hydrophilic surfaces allow graphene oxide nanosheets to have good compatibility with ethylene glycol. Yu et al. [56] used a facial technique to prepare ethylene glycol-based nanofluids containing graphene nanosheets. They found that the thermal conductivity of the graphene nanofluids increased by 86% at 5.0 vol.% graphene, showing that the extensive presence of saturated sp^3 bonds and oxygen atoms made graphene oxide nonconducting, hindered the thermal transport, and promoted phonon scattering. Mehrali et al. [57] prepared stable nanofluids by mixing graphene nanoplatelets and water using a tip ultrasonicator. In this study, four concentrations were prepared: 0.025, 0.05, 0.075, and 0.1 wt.% for three different specific surface areas of 300, 500, and 750 m^2/g. It was found that the highest thermal conductivity was 27.64% for a concentration of 0.1 wt.%, with a specific surface area of 750 m^2/g. Aravind and Ramaprabhu [58] adopted a simple thermal treatment method to prepare graphene nanosheets (GNs)-based nanofluids. The procedure was carried out without surfactant and harsh chemical treatments, which reduced the alkaline pH of graphite oxide suspension in ethylene glycol (EG) and deionized (DI) water-based fluids. It was found that the thermal conductivity at 25°C and 0.14% volume fraction of GN in EG and DI water is enhanced by 6.5 and 13.6%, respectively. Ijam et al. [59] prepared a glycerol-water-based nanofluid containing graphene oxide nanosheets. The mixture was stable for up to 5 months. The thermal conductivity of the prepared nanofluid with different temperatures (25–45°C) and weight fractions (0.02–0.1 wt.%) revealed an enhancement of thermal conductivity of 4.5% at 25°C for a weight fraction of 0.02 and 11.7% for a weight fraction of 0.1 wt.% and 45°C, respectively. Kole et al. [60] studied the thermal and electrical conductivity of stable and well dispersed functionalized graphene-ethylene glycol. High purity graphite powder was used to make Graphene nanosheets. Then the Hummers' method [61] was followed, and exfoliation and reduction by hydrogen gas were used to obtain hydrogen-exfoliated graphene (HEG), which was functionalized with acid. The material was dispersed in distilled water nanofluids with volume concentration between 0.041 and 0.395 vol.%. It was found that the thermal conductivity at 0.395 vol.% was enhanced by 15%, and the thermal conductivity of both the base fluid and the prepared nanofluid increased linearly with temperature. However, the thermal conductivity ratios (TC of the nanofluids over the TC of base fluids) were nearly independent of temperature. The study also showed that the electrical conductivity of the f-HEG nanofluids had a significant enhancement of 8620% at 0.395 vol.% loading of f-HEG, in a base fluid of 70:30 mixture of EG and distilled water. Hajjar et al. [62] studied the thermal conductivity of graphene oxide nanofluids, which comprised of graphene oxide obtained by the Hummers' method and water. They found that 0.25 wt.% of graphene oxide could enhance the thermal conductivity by 33.9% at 20°C and up to 47.5% at 40°C. Ghozatloo et al. [63] investigated the stability and thermal conductivity behavior of the nanofluids made of graphene- and water-based fluids. They also analyzed the influence of time and temperature on the effective thermal conductivities for different concentrations of graphene that was functionalized in mild conditions by potassium persulfate. The thermal conductivity for 0.05 wt.% of the functionalized graphene is enhanced around 14.1% at 25°C and 17% at 50°C compared with water. Liu et al. [64] studied the thermodynamic properties (including thermal conductivity, viscosity, specific heat, and density) of graphene-dispersed nanofluids for temperatures ranging from room temperature to around 200°C. The studied nanofluids were made out of graphene and the ionic liquid 1-hexyl-3-methylimidazolium tetra fluoroborate ((HMIM)BF4). The work showed that the thermal conductivity of the nanofluid containing 0.06 wt.% graphene increased by 15.2–22.9% at the tested temperature range (25–200°C). Viscosity, specific heat, and density decreased, respectively, by 4.6–13.1, 3, and 2.8%, compared to (HMIM)BF4.

3. Effect of the alignment of the carbon nanostructures on the thermal conductivity of nanofluids

Many researchers have studied the effect of the alignment of iron oxide nanoparticles and carbon nanomaterials on the thermal, electrical, and mechanical properties of fluids and polymers. **Table 2** summarizes most work about the effect of the alignment of iron oxide nanoparticles and carbon nanomaterials on the thermal conductivity of nanofluids.

Reference	Material	Base fluid	Concentration	Characterization	TC improvement		Stabilizing agent	Stability
					No MF	With MF		
Hong et al. [65]	CNTs +Fe_2O_3	Water	0.01 wt.% CNTs, 0.02 wt.% Fe_2O_3	SEM	15%	52%	Sodium dodecylbenzenesulfonate (NaDDBS)	Stable
Wensel et al. [17]	CNTs +Fe_2O_3	Water	0.02 wt.%	Optical microscopy	10 %	50%	NaDDBS	Stable
Wright et al. [66]	Ni-CNTs	Water	0.01 wt.%	NA	NA	75%	NA	Stable
Horton et al. [67]	Ni-CNTs	PAO	0.05 wt.%	TEM, Optical microscopy	NA	11%	NA	Stable
Hong et al. [68]	SWCNT, Fe_2O_3	Water	0.017 wt.% SWNT, 0.017 wt.% Fe_2O_3	TEM, Optical microscopy		MF 0.62 kG applied for 30 min 123%	NaDDBS	Stable
Younes et al. [69]	Fe_2O_3	Water	0.4 vol.%	TEM, Optical microscopy	18%	81%	NA	Stable

Reference	Material	Base fluid	Concentration	Characterization	TC improvement		Stabilizing agent	Stability
Younes et al. [70]	SWCNT-COOH	Water	0.016 wt.%	Optical microscopy	NA	41%	NA	Stable
	SWCNT-SO_3H	Water				40%		Stable
	SWCNT-PABS	Water				12%		Stable
	SWCNT-CONH2	Water				17%		Stable
	SWCNT-PEG	Water				14%		Stable
Younes et al. [30]	CNFs	Water	0.02 wt.%	Optical microscopy	10%	20%	NaDDBS	Stable
Philip et al. [71]	Fe_3O_4	Water	6.3 vol.%	Optical microscopy		300%	NA	Stable

Table 2. Current research of the effect of the alignment of iron oxide nanoparticles and carbon nanomaterials on the thermal conductivity of nanofluids.

Hong et al. [65] studied the effect of the alignment on the thermal conductivity of nanofluids under an external magnetic field and obtained the thermal conductivity and microscopic images taken with and without magnetic field, shown in **Figures 6** and **7**.

Figure 6. Experimental setup for taking optical microscopy images using the optical microscope, Leica Z16 APO under application of a magnetic field. From Ref. [70].

Figure 7. Experimental setup for measuring TC under application of a magnetic field. From Ref. [70].

Figure 8 shows that the nanotubes align very well in the magnetic field direction.

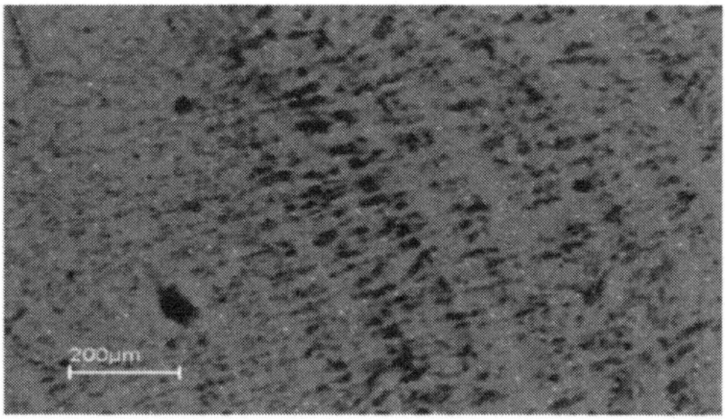

Figure 8. SEM picture of 0.01 wt.% carbon nanotube + 0.02 wt.% Fe_2O_3 with magnetic field. From Ref. [65].

The heat transfer of nanofluids containing carbon nanotubes and magnetic-field-sensitive nanoparticles of Fe_2O_3 was enhanced when a magnetic field was applied for some time. The enhancement was attributed to the formation of aligned chains of Fe_2O_3 under the applied magnetic field and to the fact that chains help to connect the nanotubes. Wensel and Wright [17, 66] found that the thermal conductivity of nanofluids containing 0.02 wt.% was enhanced by 10% under a magnetic field. That was attributed to the aggregation of metal oxide particles on the surface of nanotubes by electrostatic attraction, which forms the aggregation chain along the nanotubes. Horton et al. [67] investigated the thermal conductivity of nanofluids containing magnetic-metal-coated carbon nanotubes (Ni-coated SWCNTs), which significantly enhanced by 60% under a magnetic field. **Figure 9** shows a color online microscope image of 0.05 wt.% Ni-coated SWCNT in DI water, before (A) and after the magnetic field (B) was applied ($H = 0.62$ kG).

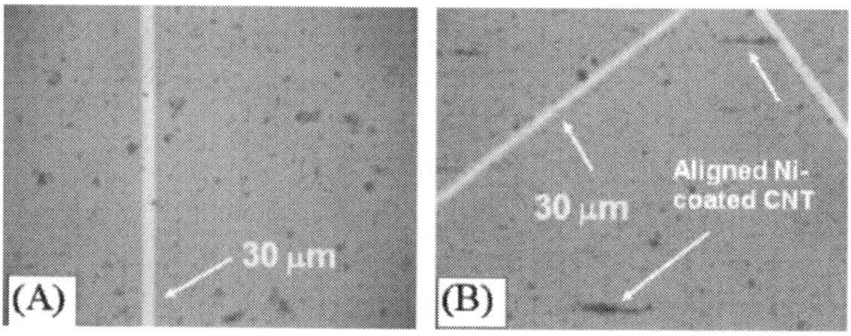

Figure 9. Color online microscope image of 0.05 wt.% Ni-coated SWCNT in DI water. (A) Before magnetic field and (B) after magnetic field ($H = 0.62$ kG). From Ref. [66].

Figure 10 shows TEM images of carbon nanotubes. (A) As received, uncoated CNTs, the image shows that the wall of one CNT is clear and nothing adhere to it. (B) Ni-coated CNTs, the image shows that the wall of the CNTs has been modified with some Ni adhered to the surface of the CNTs and (C) shows that the wall of the Ni-coated CNTs after applying the magnetic field experiment. It can be seen that the density of the Ni coated has decreased due to the effect of sonication and the magnetic field experiment.

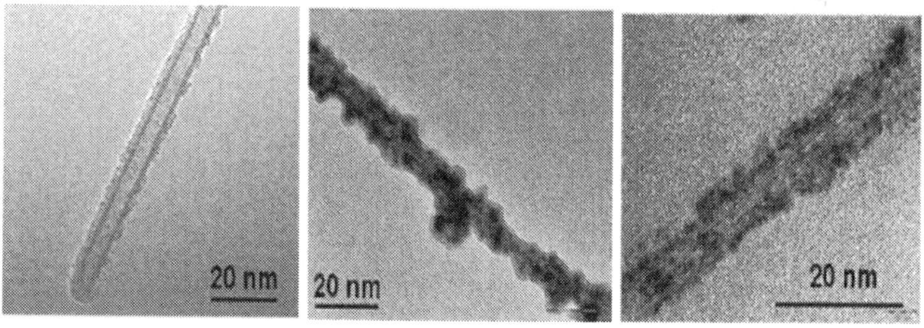

Figure 10. TEM images of carbon nanotubes. (A) As received, uncoated. (B) Ni-coated nanotube before the magnetic field experiment, and (C) after the experiment. From Ref [67].

Hong et al. [68] studied the thermal conductivity of nanofluids containing CNTs and magnetic-field-sensitive nanoparticles of Fe_2O_3 with NaDDBS surfactant and found that an electrostatic attraction among the nanotubes, the surfactant, and the metal oxide caused aggregation. NaDDBS surfactant with a negative charge attracted Fe_2O_3, which had positive zeta potential charge. When a cationic surfactant, cetyltrimethyl ammonium bromide (CTAB), replaced the anionic surfactant NaDDBS, no alignment was found because both the surfactant and Fe_2O_3 have the same positive charge. **Figure 11** shows that the nanotubes align very well in the direction of the magnetic field, either in scale bar 100 μm (A) or 10 μm (B).

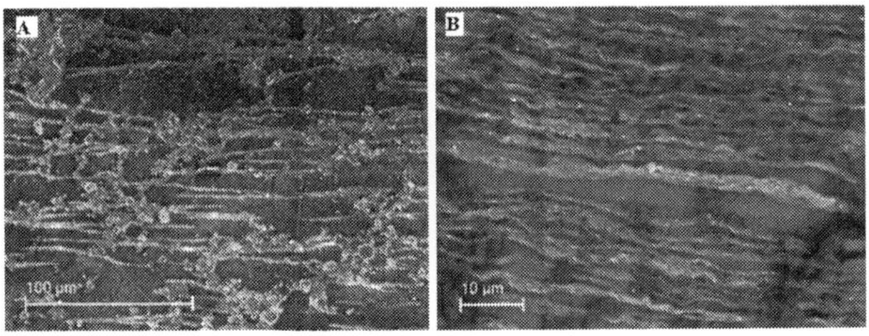

Figure 11. Electron SEM images of 0.017 wt.% SWCNT, 0.017 wt.% Fe_2O_3, and 0.17 wt.% NaDSSB with magnetic field. (A) Scale bar 100 m, and (B) scale bar 10 m. From Ref. [68].

Younes et al. [27, 70] studied the alignment of different functionalized SWCNTs using Fe_2O_3 nanoparticles under an external magnetic field. These authors found that, even in the absence of NaDDBS as surfactant, some functionalized SWNTs, i.e., SWNT-SO_2OH and SWNT-COOH, could disperse well in water and showed a clear alignment under an external magnetic field. **Figure 12** shows microscopy image of 0.016 wt.% SWNT-COOH and 0.016 wt.% Fe_2O_3 in DI water. This microscope image of 0.016 wt.% SWNT-COOH and 0.016 wt.% γ-Fe_2O_3 in DI water was obtained using a high-speed microscope video system. Magnetic field (H = 0.14 kG) was applied with an internal reference of 30 μm. As shown in **Figure 12**, it is clearly seen that the randomly dispersed SWNT-COOH/γ-Fe_2O_3 forms large and long lines, indicating that SWNT-COOH/γ-Fe_2O_3 nanoparticle mixture aligns under the external magnetic field.

Figure 12. Microscopy image of 0.016 wt.% SWNT-COOH, 0.016 wt.% Fe_2O_3 in DI water. At 8 min, the internal reference is 30 μm. From Ref. [70].

Younes et al. [30] studied the effect of the alignment of carbon nanofibers in water and epoxy under an external magnetic field on the thermal conductivity of nanofluids. They found that the magnetic field aligned the nanofibers, not only in water but also in epoxy, as clearly seen in **Figure 13**. Adding 0.02 wt.% of carbon nanofibers to water enhanced the thermal conductivity by 10%. The addition of an external magnetic field enhanced the thermal conductivity around 20%.

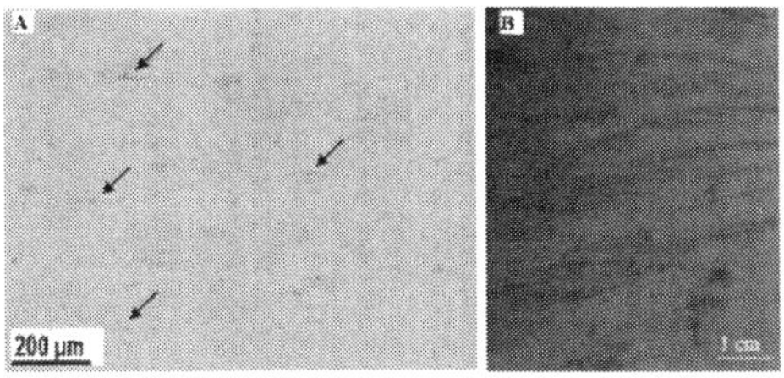

Figure 13. Alignment of CNFs in epoxy. (A) Optical microscopy images for 0.002 wt.% CNFs, 0.002 wt.% Fe_2O_3, and 0.02 wt.% NaDDBS in epoxy after 48 h. (B) Digital camera image of 0.02 wt.% CNFs, 0.02 wt.% Fe_2O_3, and 0.2 wt.% NaDDBS in epoxy after 2 h. Magnetic field (H = 1.2 kG) was applied. From Ref. [30].

Christensen et al. [26] studied the effect of nonionic surfactant on the dispersion and alignment of carbon nanotubes in the presence of Fe_2O_3. Gum arabic and Triton X-100 were used as nonionic surfactants. Gum arabic did not allow good dispersion or alignment in aqueous fluid. Triton X-100 allowed dispersion and alignment of both Fe_2O_3 and SWNT in an aqueous base

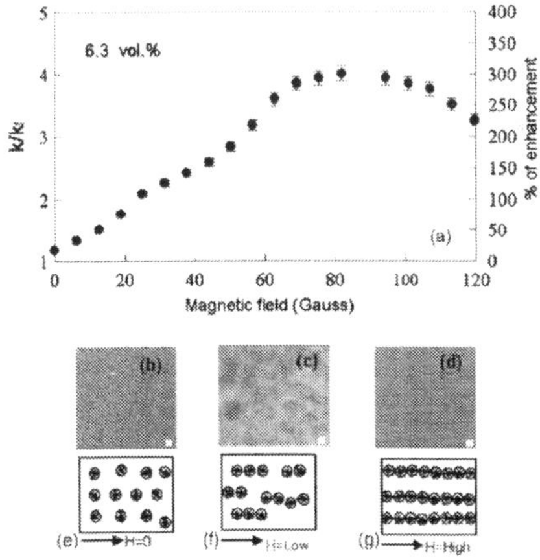

Figure 14. (a) Thermal conductivity ratio k/k_f and the corresponding percentage of enhancement as a function of applied magnetic field for 6.3 vol.% of Fe_3O_4. (B)–(D) Micrographs of ferrofluid emulsion of 200 nm, (B) without, (C) with low, and (D) with high magnetic field. (E)–(G) Schematic alignment of particles under the above three cases. From Ref. [71].

fluid. Due to the partial negative charge of Triton X-100, alignment did occur, but the alignment was slower, when compared to surfactants or functional groups containing stronger charges.

Glover et al. [72] prepared a nanofluids that consisted of water and 0–0.5 wt.% of sulfonated carbon nanotubes. The electrical conductivity of the prepared nanofluids was measured. The study found that the electrical conductivity increased by 13 times when only 0.5 wt.% of functional sulfonated carbon nanotubes was used. Philip et al. [71] observed a dramatic enhancement of thermal conductivity in a nanofluid containing magnetic particles of Fe_3O_4 (of 6.7 nm diameter) under a magnetic field (see **Figure 14**). The maximum enhancement in thermal conductivity was 300% ($k/k_f = 4.0$) for a particle loading of 6.3 vol.%.

4. Conclusions

The main focus of this chapter is to summarize the recent studies using CNTs and graphene to improve the thermal conductivity of nanofluids. Moreover, this chapter summarizes the investigation about the effect of using magnetic fields on enhancing the thermal conductivity of nanofluids. Too much discrepancy about thermal conductivity of nanofluids can be found in the literature. For carbon nanofluids, unfortunately, no significant improvements on thermal conductivity are observed using low concentrations. The thermal conductivity increased by approximately 10–20% when 1 vol.% SWCNT was used. However, at these high concentrations, the viscosity is greatly increased and the nanofluid becomes "mud-like," which are the difficulties reported in the literature. This variation in the thermal conductivity can be attributed to many factors, such as particle size, temperature, pH, or zeta potential. Finally, we believe that more efforts need to be done in order to improve the thermal conductivity of nanofluids and study the effect of the different parameters in thermal conductivity. In addition, more investigation about the use of magnetic field to align magnetic sensitive nanoparticles is required.

Author details

Hammad Younes[1*], Amal Al Ghaferi[1], Irfan Saadat[2] and Haiping Hong[3]

*Address all correspondence to: hyounes@masdar.ac.ae

1 Department of Mechanical and Material Engineering, Masdar Institute of Science and Technology, Abu Dhabi, United Arab Emirates

2 Department of Electrical Engineering and Computer Science, Masdar Institute of Science and Technology, Abu Dhabi, United Arab Emirates

3 Department of Material and Metallurgical Engineering, South Dakota School of Mines and Technology, Rapid City, SD, USA

References

[1] Altan CL, Elkatmis A, Yüksel M, Aslan N, Bucak S. Enhancement of thermal conductivity upon application of magnetic field to Fe3O4 nanofluids. Journal of Applied Physics. 2011;110(9):093917.

[2] Leong K, Saidur R, Kazi S, Mamun A. Performance investigation of an automotive car radiator operated with nanofluid-based coolants (nanofluid as a coolant in a radiator). Applied Thermal Engineering. 2010;30(17):2685–2692.

[3] Srikant R, Rao D, Subrahmanyam M, Krishna VP. Applicability of cutting fluids with nanoparticle inclusion as coolants in machining. Proceedings of the Institution of Mechanical Engineers, part J. Journal of Engineering Tribology. 2009;223(2):221–225.

[4] Kulkarni DP, Vajjha RS, Das DK, Oliva D. Application of aluminum oxide nanofluids in diesel electric generator as jacket water coolant. Applied Thermal Engineering. 2008;28(14):1774–1781.

[5] K.K. Kuo, G.A. Risha, B.J. Evans, E. Boyer, Potential Usage of Energetic Nano-sized Powders for Combustion and Rocket Propulsion, MRS Online Proceedings Library Archive, 800 (2003) AA1.1 (12 pages).

[6] Prasher R, Phelan PE, Bhattacharya P. Effect of aggregation kinetics on the thermal conductivity of nanoscale colloidal solutions (nanofluid). Nano Letters. 2006;6(7):1529–1534.

[7] Wen D, Lin G, Vafaei S, Zhang K. Review of nanofluids for heat transfer applications. Particuology. 2009;7(2):141–150.

[8] Saidur R, Leong K, Mohammad H. A review on applications and challenges of nanofluids. Renewable and Sustainable Energy Reviews. 2011;15(3):1646–1668.

[9] Pastoriza-Gallego M, Lugo L, Legido J, Piñeiro M. Enhancement of thermal conductivity and volumetric behavior of FexOy nanofluids. Journal of Applied Physics. 2011;110(1):014309.

[10] Yu W, France DM, Routbort JL, Choi SU. Review and comparison of nanofluid thermal conductivity and heat transfer enhancements. Heat Transfer Engineering. 2008;29(5): 432–460.

[11] Sidik NAC, Mohammed H, Alawi OA, Samion S. A review on preparation methods and challenges of nanofluids. International Communications in Heat and Mass Transfer. 2014;54:115–125.

[12] Choi S, Zhang Z, Yu W, Lockwood F, Grulke E. Anomalous thermal conductivity enhancement in nanotube suspensions. Applied Physics Letters. 2001;79(14):2252–2254.

[13] Keblinski P, Eastman JA, Cahill DG. Nanofluids for thermal transport. Materials Today. 2005;8(6):36–44.

[14] Wang X, Xu X, Choi SU. Thermal conductivity of nanoparticle-fluid mixture. Journal of Thermophysics and Heat Transfer. 1999;13(4):474–480.

[15] Wen D, Ding Y. Experimental investigation into the pool boiling heat transfer of aqueous based γ-alumina nanofluids. Journal of Nanoparticle Research. 2005;7(2–3): 265–274.

[16] Anoop K, Sundararajan T, Das SK. Effect of particle size on the convective heat transfer in nanofluid in the developing region. International Journal of Heat and Mass Transfer. 2009;52(9):2189–2195.

[17] Wensel J, Wright B, Thomas D, Douglas W, Mannhalter B, Cross W, et al. Enhanced thermal conductivity by aggregation in heat transfer nanofluids containing metal oxide nanoparticles and carbon nanotubes. Applied Physics Letters. 2008;92(2):023110.

[18] Iijima S. Helical microtubules of graphitic carbon. Nature. 1991;354(6348):56–58.

[19] Biercuk M, Llaguno MC, Radosavljevic M, Hyun J, Johnson AT, Fischer JE. Carbon nanotube composites for thermal management. Applied Physics Letters. 2002;80(15): 2767–2769.

[20] Berber S, Kwon Y-K, Tománek D. Unusually high thermal conductivity of carbon nanotubes. Physical Review Letters. 2000;84(20):4613.

[21] Nan C-W, Liu G, Lin Y, Li M. Interface effect on thermal conductivity of carbon nanotube composites. Applied Physics Letters. 2004;85(16):3549–3551.

[22] Spitalsky Z, Tasis D, Papagelis K, Galiotis C. Carbon nanotube–polymer composites: chemistry, processing, mechanical and electrical properties. Progress in Polymer Science. 2010;35(3):357–401.

[23] Hongtao L, Hongmin J, Haiping H, Younes H. Tribological properties of carbon nanotube grease. Industrial Lubrication and Tribology. 2014;66(5):579–583.

[24] Abueidda DW, Dalaq AS, Al-Rub RKA, Younes HA. Finite element predictions of effective multifunctional properties of interpenetrating phase composites with novel triply periodic solid shell architectured reinforcements. International Journal of Mechanical Sciences. 2015;92:80–89.

[25] Abueidda DW, Al-Rub RKA, Dalaq AS, Younes HA, Al Ghaferi AA, Shah TK. Electrical conductivity of 3D periodic architectured interpenetrating phase composites with carbon nanostructured-epoxy reinforcements. Composites Science and Technology. 2015;118:127–134.

[26] Christensen G, Younes H, Hong H, Smith P. Effects of solvent hydrogen bonding, viscosity, and polarity on the dispersion and alignment of nanofluids containing Fe_2O_3 nanoparticles. Journal of Applied Physics. 2015;118(21):214302.

[27] H. Younes, Carbon Nanomaterials and Metal Oxide Nanoparticles and Their Applications Toward Grease and Nanofluids, South Dakota School of Mines and Technology, Rapid City2013.

[28] Sofela SO, Younes H, Jelbuldina M, Saadat I, Al Ghaferi A, editors. Carbon nanomaterials based TSVs for dual sensing and vertical interconnect application. Interconnect Technology Conference and 2015 IEEE Materials for Advanced Metallization Conference (IITC/MAM), 2015 IEEE International; 2015.

[29] Younes H, Christensen G, Li D, Hong H, Ghaferi AA. Thermal conductivity of nanofluids: Review. Journal of Nanofluids. 2015;4(2):107–132.

[30] Younes H, Christensen G, Liu M, Hong H, Yang Q, Lin Z. Alignment of carbon nanofibers in water and epoxy by external magnetic field. Journal of Nanofluids. 2014;3(1):33–37.

[31] Rahmana M, Younes H, Ni G, Zhang T, Al Ghaferi A. Synthesis and optical characterization of carbon nanotube arrays. Materials Research Bulletin. 2016;77:243–252.

[32] Rahman MM, Younes H, Subramanian N, Ghaferi AA. Optimizing the dispersion conditions of SWCNTs in aqueous solution of surfactants and organic solvents. Journal of Nanomaterials. 2014;2014:145.

[33] Cui S, Canet R, Derre A, Couzi M, Delhaes P. Characterization of multiwall carbon nanotubes and influence of surfactant in the nanocomposite processing. Carbon. 2003;41(4):797–809.

[34] Yurekli K, Mitchell CA, Krishnamoorti R. Small-angle neutron scattering from surfactant-assisted aqueous dispersions of carbon nanotubes. Journal of the American Chemical Society. 2004;126(32):9902–9903.

[35] Rosca ID, Watari F, Uo M, Akasaka T. Oxidation of multiwalled carbon nanotubes by nitric acid. Carbon. 2005;43(15):3124–3131.

[36] Velasco-Santos C, Martínez-Hernández AL, Fisher FT, Ruoff R, Castano VM. Improvement of thermal and mechanical properties of carbon nanotube composites through chemical functionalization. Chemistry of Materials. 2003;15(23):4470–4475.

[37] Saini RK, Chiang IW, Peng H, Smalley R, Billups W, Hauge RH, et al. Covalent sidewall functionalization of single wall carbon nanotubes. Journal of the American Chemical Society. 2003;125(12):3617–3621.

[38] Kong H, Gao C, Yan D. Functionalization of multiwalled carbon nanotubes by atom transfer radical polymerization and defunctionalization of the products. Macromolecules. 2004;37(11):4022–4030.

[39] Peng H, Alemany LB, Margrave JL, Khabashesku VN. Sidewall carboxylic acid functionalization of single-walled carbon nanotubes. Journal of the American Chemical Society. 2003;125(49):15174–15182.

[40] Shen J, Huang W, Wu L, Hu Y, Ye M. The reinforcement role of different amino-functionalized multi-walled carbon nanotubes in epoxy nanocomposites. Composites Science and Technology. 2007;67(15):3041–3050.

[41] Jeong G-H. Surface functionalization of single-walled carbon nanotubes using metal nanoparticles. Transactions of Nonferrous Metals Society of China. 2009;19(4):1009–1012.

[42] Zhang L, Kiny VU, Peng H, Zhu J, Lobo RF, Margrave JL, et al. Sidewall functionalization of single-walled carbon nanotubes with hydroxyl group-terminated moieties. Chemistry of Materials. 2004;16(11):2055–2061.

[43] Phuoc TX, Massoudi M, Chen R-H. Viscosity and thermal conductivity of nanofluids containing multi-walled carbon nanotubes stabilized by chitosan. International Journal of Thermal Sciences. 2011;50(1):12–18.

[44] Garg P, Alvarado JL, Marsh C, Carlson TA, Kessler DA, Annamalai K. An experimental study on the effect of ultrasonication on viscosity and heat transfer performance of multi-wall carbon nanotube-based aqueous nanofluids. International Journal of Heat and Mass Transfer. 2009;52(21–22):5090–5101.

[45] Meng Z, Wu D, Wang L, Zhu H, Li Q. Carbon nanotube glycol nanofluids: photo-thermal properties, thermal conductivities and rheological behavior. Particuology. 2012;10(5):614–618.

[46] Chen L, Xie H, Li Y, Yu W. Nanofluids containing carbon nanotubes treated by mechanochemical reaction. Thermochimica Acta. 2008;477(1–2):21–24.

[47] Chen L, Xie H. Surfactant-free nanofluids containing double- and single-walled carbon nanotubes functionalized by a wet-mechanochemical reaction. Thermochimica Acta. 2010;497(1–2):67–71.

[48] Assael M, Metaxa I, Arvanitidis J, Christofilos D, Lioutas C. Thermal conductivity enhancement in aqueous suspensions of carbon multi-walled and double-walled nanotubes in the presence of two different dispersants. International Journal of Thermophysics. 2005;26(3):647–664.

[49] Ding Y, Alias H, Wen D, Williams RA. Heat transfer of aqueous suspensions of carbon nanotubes (CNT nanofluids). International Journal of Heat and Mass Transfer. 2006;49(1–2):240–250.

[50] Shaikh S, Lafdi K, Ponnappan R. Thermal conductivity improvement in carbon nanoparticle doped PAO oil: An experimental study. Journal of Applied Physics. 2007;101(6):064302.

[51] Theres Baby T, Sundara R. Synthesis of silver nanoparticle decorated multiwalled carbon nanotubes-graphene mixture and its heat transfer studies in nanofluid. AIP Advances. 2013;3(1):012111.

[52] Xie H, Lee H, Youn W, Choi M. Nanofluids containing multiwalled carbon nanotubes and their enhanced thermal conductivities. Journal of Applied Physics. 2003;94(8):4967.

[53] Novoselov KS, Geim AK, Morozov SV, Jiang D, Zhang Y, Dubonos SV, Grigorieva IV, Firsov AA. Electric field effect in atomically thin carbon films. Science. 2004; 306(5696): 666–669.

[54] Cooper DR, D'Anjou B, Ghattamaneni N, Harack B, Hilke M, Horth A, Majlis N, Massicotte M, Vandsburger L, Whiteway E, Yu V. Experimental review of graphene. ISRN Condensed Matter Physics. 2011;2012:1–56.

[55] Yu W, Xie H, Bao D. Enhanced thermal conductivities of nanofluids containing graphene oxide nanosheets. Nanotechnology. 2009;21(5):055705.

[56] Yu W, Xie H, Wang X, Wang X. Significant thermal conductivity enhancement for nanofluids containing graphene nanosheets. Physics Letters A. 2011;375(10):1323–1328.

[57] Mehrali M, Sadeghinezhad E, Latibari ST, Kazi SN, Mehrali M, Zubir MNBM, et al. Investigation of thermal conductivity and rheological properties of nanofluids containing graphene nanoplatelets. Nanoscale Research Letters. 2014;9(1):1–12.

[58] Aravind SJ, Ramaprabhu S. Surfactant free graphene nanosheets based nanofluids by in-situ reduction of alkaline graphite oxide suspensions. Journal of Applied Physics. 2011;110(12):124326.

[59] Ijam A, Golsheikh AM, Saidur R, Ganesan P. A glycerol–water-based nanofluid containing graphene oxide nanosheets. Journal of Materials Science. 2014;49(17):5934–5944.

[60] Kole M, Dey T. Investigation of thermal conductivity, viscosity, and electrical conductivity of graphene based nanofluids. Journal of Applied Physics. 2013;113(8):084307.

[61] Hummers Jr WS, Offeman RE. Preparation of graphitic oxide. Journal of the American Chemical Society. 1958;80(6):1339.

[62] Hajjar Z, Morad Rashidi A, Ghozatloo A. Enhanced thermal conductivities of graphene oxide nanofluids. International Communications in Heat and Mass Transfer. 2014;57:128–131.

[63] Ghozatloo A, Shariaty-Niasar M, Rashidi AM. Preparation of nanofluids from functionalized graphene by new alkaline method and study on the thermal conductivity and stability. International Communications in Heat and Mass Transfer. 2013;42:89–94.

[64] Liu J, Wang F, Zhang L, Fang X, Zhang Z. Thermodynamic properties and thermal stability of ionic liquid-based nanofluids containing graphene as advanced heat transfer fluids for medium-to-high-temperature applications. Renewable Energy. 2014;63:519–523.

[65] Hong H, Wright B, Wensel J, Jin S, Ye XR, Roy W. Enhanced thermal conductivity by the magnetic field in heat transfer nanofluids containing carbon nanotube. Synthetic Metals. 2007;157(10):437–440.

[66] Wright B, Thomas D, Hong H, Groven L, Puszynski J, Duke E, et al. Magnetic field enhanced thermal conductivity in heat transfer nanofluids containing Ni coated single wall carbon nanotubes. Applied Physics Letters. 2007;91(17):173116.

[67] Horton M, Hong H, Li C, Shi B, Peterson G, Jin S. Magnetic alignment of Ni-coated single wall carbon nanotubes in heat transfer nanofluids. Journal of Applied Physics. 2010;107(10):104320.

[68] Hong H, Luan X, Horton M, Li C, Peterson G. Alignment of carbon nanotubes comprising magnetically sensitive metal oxides in heat transfer nanofluids. Thermochimica Acta. 2011;525(1):87–92.

[69] Younes H, Christensen G, Luan X, Hong H, Smith P. Effects of alignment, pH, surfactant, and solvent on heat transfer nanofluids containing Fe2O3 and CuO nanoparticles. Journal of Applied Physics. 2012;111(6):064308.

[70] Younes H, Christensen G, Hong H, Peterson G. Alignment of different functionalized single wall carbon nanotubes using Fe2O3 nanoparticles under external magnetic field. Journal of Nanofluids. 2013;2(1):4–10.

[71] Philip J, Shima P, Raj B. Enhancement of thermal conductivity in magnetite based nanofluid due to chainlike structures. Applied Physics Letters. 2007;91(20):203108.

[72] Glover B, Whites KW, Hong H, Mukherjee A, Billups WE. Effective electrical conductivity of functional single-wall carbon nanotubes in aqueous fluids. Synthetic Metals. 2008;158(12):506–508.

Chapter 10

Carbon Nanostructure-Based Scale Sensors Using Inkjet Printing and Casting Techniques

Hammad Younes, Amal Al Ghaferi and Irfan Saadat

Additional information is available at the end of the chapter

http://dx.doi.org/10.5772/64196

Abstract

In this chapter, the fabrication and characterization of scale sensor using carbon nanotubes (CNTs) are discussed. Two different methods are used to prepare the carbon nanomaterials for the sensor fabrication: CNT casting and the CNT inkjet printing. In addition, the sensors are integrated into Kelvin architectures. The electrical resistance of the carbon nanomaterial films is measured with and without adding a drop of brine to the surface of the film. The films are characterized by scanning electron microscopy (SEM), atomic force microscopy (AFM), transmission electron microscopy (TEM), and energy dispersive X-ray spectroscopy (EDS). Electrical resistance of the casted CNT films and five layers of CNT inkjet printing are found to be close to 40.0 kΩ and 1.00 kΩ, respectively. Adding one drop of brine solution on the surface of the casted CNT film and five layers of CNT inkjet printing changed the resistance by 50% and 75%, respectively. The resetting process is done for all sensors by soaking in deionized water (DI water) for some time, and the electrical resistance is measured and found to be close to the initial electrical resistance.

Keywords: CNTs, Sensors, inkjet printing, smear casting, scale and electrical resistance

1. Introduction

Carbon nanotubes (CNTs) are an important new class of nano-2-D and nano-1-D materials that have numerous novel and useful properties. Their nanostructures exhibit many interesting and often unique properties, and hence, they can be used in many novel technological and industrial applications. Due to their unique electrical and electronic properties, CNTs are considered for various next-generation device applications, that is nanofluids [1–3], scanning probes

[4], nanocomposites [5–9], grease [10, 11], electron magnetic shielding interface [12–16], solar thermal absorbers [17] and sensors [18–22]. The electrical resistance of the carbon nanomaterials changes when chemicals in the surrounding make covalent or non-covalent interaction with the carbon nanomaterials. This change makes carbon nanomaterials good candidates for chemical sensors. The absorbed molecules act as dopants that cause shifting on the Fermi energy of the carbon nanomaterials. In addition, the bonds formed between absorbed chemicals and the carbon nanomaterials change the band structure of the tube [23].

Scale is the creation of inorganic soluble salts from aqueous brines during oil and gas production, which deposits at the internal surface of the pipelines under the supersaturating conditions [24]. Calcium carbonate ($CaCO_3$) and barium sulfate ($BaSO_4$) are the most insoluble types of inorganic scale formed during the extraction of oil. The deposition can be seen not only in the pipelines but also in water-handling equipment, pumps, valves, and any other parts that interact with water. The deposition of the scale leads to unavoidable damage of the equipment parts. Therefore, suspension of oil operations becomes essential in order to replace the damaged parts. In the petroleum industry, such interruptions lead to extremely high costs [25].

There has been multiple studies and research conducted on scale monitoring, testing, and optimization of inhibitors [26–28]. In general, these studies focused on the development of using new sensor materials that enabled monitoring and offline testing [29–32].

This chapter discusses the sensitivity and selectivity of using carbon nanotubes through two different preparation methods: the smear casted and the inkjet printing, to create sensors that detect scale in oil and gas pipelines. Moreover, the design and the microfabrication method of the sensors are discussed.

2. Experimental

Single-wall carbon nanotubes (SWCNT) were purchased from Sigma–Aldrich, Inc., USA. The CNT inkjet "CNTRENE® 3015 A3-R" used for this printing was purchased from Brewer Science, Inc., USA. DMP2831 Inkjet Printer was used to print the CNT inkjet. The chemical surfactant sodium dodecyl benzene sulfonate (NaDDBS) was purchased from Sigma–Aldrich. Epoxy resins and hardener were purchased from West System Inc, Zayed port, Abu Dhabi. Releasing agents, Frekote, were purchased from Logistics Company Limited, in Dubai, UAE. Sonication was performed using a Branson Digital Sonifier, model 450. Scanning electron microscopy (SEM) images were acquired using the backscattered electron detector on a Zeiss Supra40VP variable pressure system. The electrical conductivity was obtained using Elite 300 Semi-automatic Probe Station & Keithley 4200-SCS Parameter Analyzer. The microstructures of the samples were probed by scanning electron FEI quanta 250 ESEM and the FEI Nova NanoSEM 650. Atomic force microscopy data were obtained using (Asylum MFP-3D, AFM) deflection 1.5 V, scan speed 0.7 Hz, applied voltage 500 V, contact mode imaging.

Casted technique: The epoxy used was a pre-polymer consisting of bisphenol A attached to an epoxide group. Sodium dodecylbenzenesulfonate (NaDDBS) surfactant was added to epoxy with a weight ratio of four times carbon nanotubes and mixed manually for 5 min. The mixture was then sonicated until the surfactant was dissolved completely, and a clear solution was obtained. About 0.5% of carbon nanotubes were added to the mixture and sonicated again. The mixture was then mixed manually for 5 min with hardener by weight ratio of 10:2.63. Glass plate was coated with a layer of releasing agent (Frekote), and the mixture was casted on the glass plate by smear casting technique. The sample was cured at room temperature for 24 hours.

Carbon nanotube inkjet printing: CNT inkjet was used as it is, but the printing recipe was optimized in order to have a uniform continuous printing. CNT inkjet was printed on glass/ silicon oxide substrates.

3. Results and discussion

3.1. CNT casting

3.1.1. Preparation of carbon nanotube-based scale sensor

As it has been discussed earlier, two types of preparation techniques for the CNTs have been used: smear casted and inkjet printing. The smear casting CNT nanocomposites were prepared by dispersing 0.5 wt% of CNTs in epoxy polymer using ultrahigh tip sonicator. Kelvin structure was then fabricated as can be seen in **Figure 1**.

Figure 1. Kelvin structure for sensor device.

3.1.2. Characterization of carbon nanotube-based scale sensor

Figure 2 is a cross-sectional SEM image of the nanocomposites and shows that CNTs disperse very well in the epoxy polymer and make network structure essential for electrical conductivity. The electrical resistance of the prepared CNTs was measured just after the preparation and defined as resistance of original samples. The electrical resistance for samples were measured again after adding brine and defined as resistance with brine.

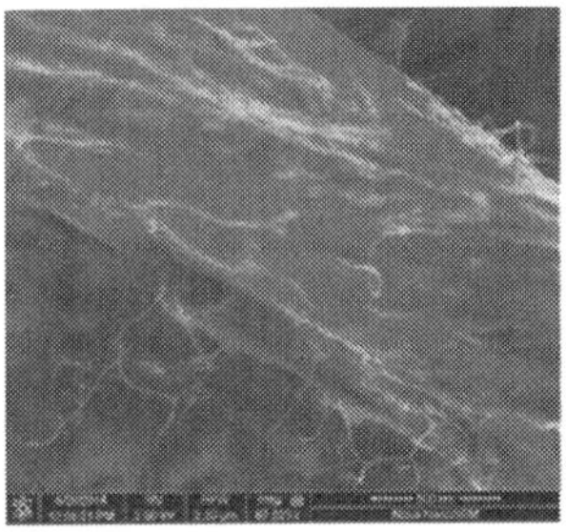

Figure 2. Images of cross section of the 1.0 wt% of CNTs in epoxy.

Figure 3 is an SEM image of the surface of smear casted CNTs. There are many white spots on the surface of the sample, so we selected one of the largest (marked with a red cross) to carry on the EDS, as shown in **Figure 4**. As it can be seen, the sample has sodium and chlorine, which indicates the presence of sodium chloride on the surface.

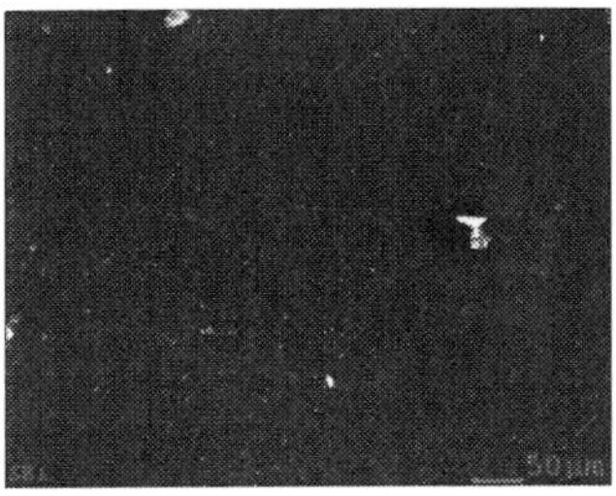

Figure 3. SEM image of the surface of smear casted CNTs showing a spot of NaCl.

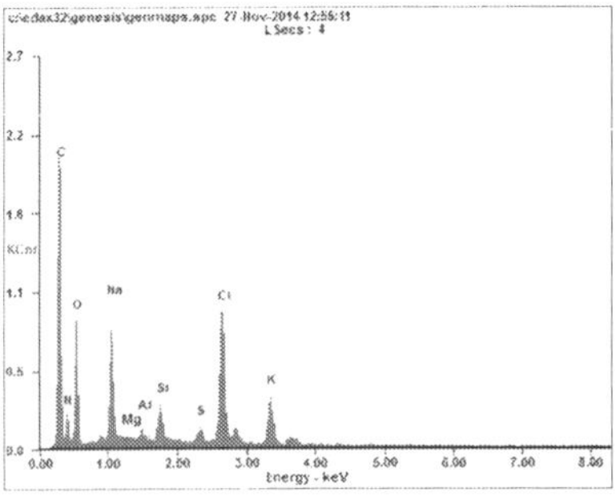

Figure 4. EDS spectrum for casted CNTs on a glass substrate.

http://dx.doi.org/10.5772/64196

Conductive atomic force microscopy (C-AFM) was carried out for sample A, which is original, and for sample B that has brine. **Figure 5** shows that sample A has more tubes than B, which means that C-AFM of casted CNTs affirm that the material A is more conductive than B, and this matches the resistance curve depicted in **Figure 6**.

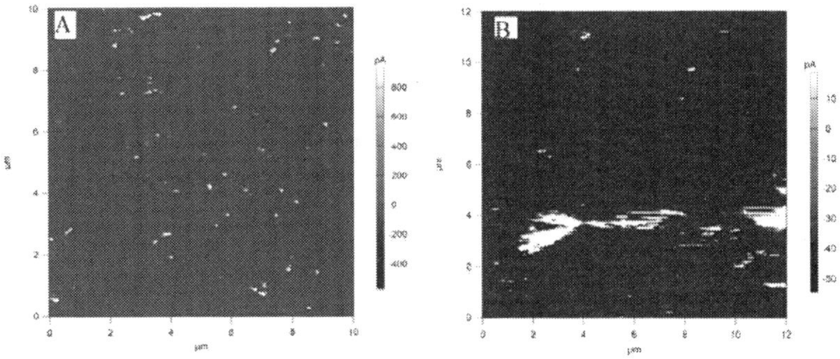

Figure 5. CAFM for current mapping of rough casted CNTs, A: original sample B: with brine sample.

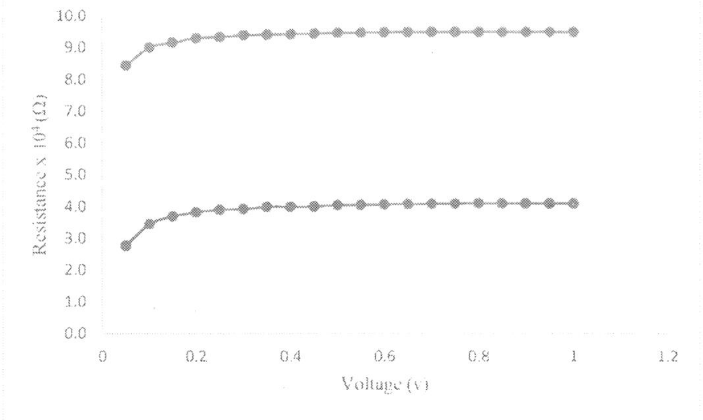

Figure 6. Electrical resistance of 0.5 wt% of smear casted CNTs.

3.1.3. *Electrical resistance of carbon nanotube-based scale sensor*

Figure 6 shows the resistance of the casted sample with sweeping voltage from 0 to 1.0. The resistance of the material seems to be very consistent with the swept voltage. Once one drop of brine solution was added to the surface of the smear casted CNTs, the resistance increased from 4.0 E+4 to almost 9.5 E+4 Ω. Chemical doping can induce strong changes in conductance. Such modifications can be easily detected by electron current signals, and these properties make CNTs extremely small sensors that are very sensitive to the surrounding chemicals.

3.2. CNT inkjet printing

3.2.1. Preparation of carbon nanotube-based scale sensor

The CNT inkjet "CNTRENE® 3015 A3-R" used for this printing was purchased from Brewer Science, Inc., USA. DMP2831 Inkjet Printer was used to print the CNT inkjet. About 1×1 cm^2 of CNT inkjet printing was printed on Si substrate. Ebeam evaporator was used to deposit 200 nm Al as metal contact using a shadow mask. The printing was carried out at different temperatures and a second curing cycle was applied to the printed samples to ensure a very staple printing.

3.2.2. Characterization of carbon nanotube-based scale sensor

After optimizing the printing recipe, it was critical to study the effect of the CNT printed layers on the electrical resistance; hereafter the sensor performance was assessed. About 3, 4, 5, and 6 layers of CNT inkjet were printed on Si substrates. The samples were characterized by measuring the electrical resistance of 4-point Kelvin-like structures and taking SEM and AFM images. **Figure 7** shows that 3, 4, 5, and 6 layers of CNT have the following electrical resistance, respectively: 5.5 kΩ, 4.2 kΩ, 1 kΩ, and 0.6 kΩ. This led us to conclude that while increasing number of printing passes, the electrical resistance decreases because the addition of CNT passes generates more paths for the electrons to go through and thus increases the conductivity.

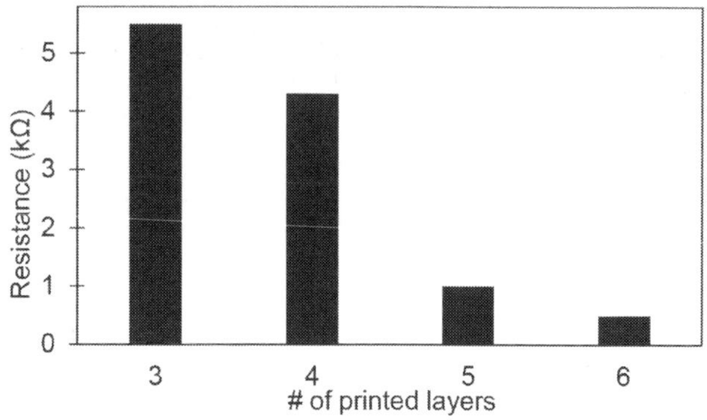

Figure 7. Electrical resistance as a function of voltage of multilayer CNT inkjet printing.

Figures 8 and **9** show SEM images for CNT inkjet printing of one and five layers of printing. The SEM images show that the CNTs are making network structures. SEM images of the five layers of CNT printing (**Figure 8**) shows more CNTs and thicker network structure than that of one layer of CNT printing (**Figure 7**), as expected.

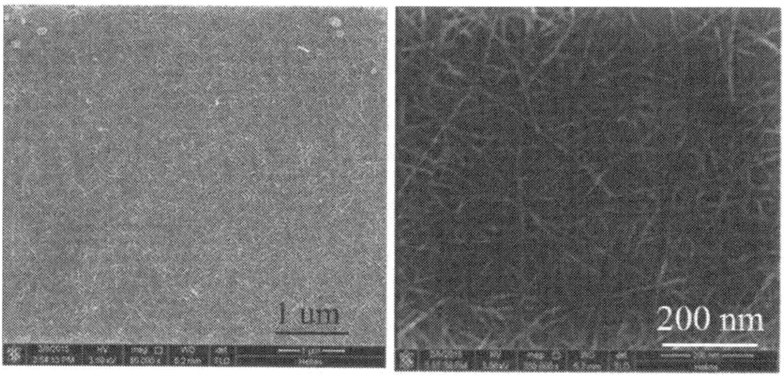

Figure 8. SEM images for one layer of CNT inkjet printed on Si substrate.

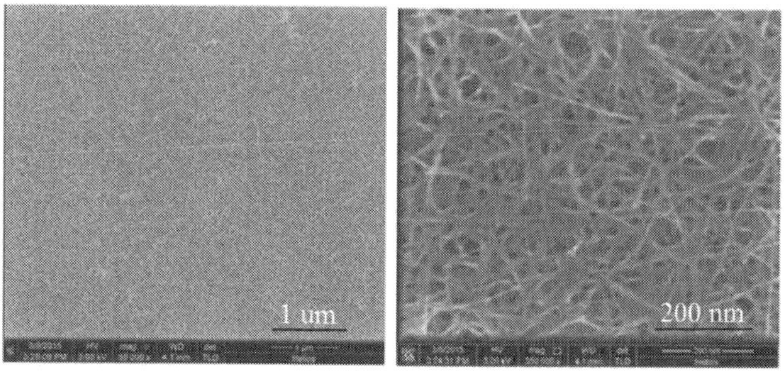

Figure 9. SEM images for five layers of CNT inkjet printed on Si substrate.

The ultimate goal of this study was to fabricate a sensor that can be used in oil and gas pipelines, which is a very hostile environment. Therefore, the sensor should be stable enough to sustain consistent results within the harsh conditions found in the pipelines. The stability of the CNT inkjet printed on silicon substrate was tested by immersing it in a solvent for some time and then obtaining SEM images, to examine the effect of the solvent on the printing.

Images obtained using tapping-mode AFM (**Figures 10** and **11**) indicate that exposure to water for 72 hours does not have pronounced effects on the topography of the sample, in the case of inkjet-printed CNT films. The roughness of the inkjet-printed CNT decreased as a result of the exposure (**Figure 8**) by 26.3%. The roughness analyses were carried out on images obtained using tapping-mode AFM topography scans.

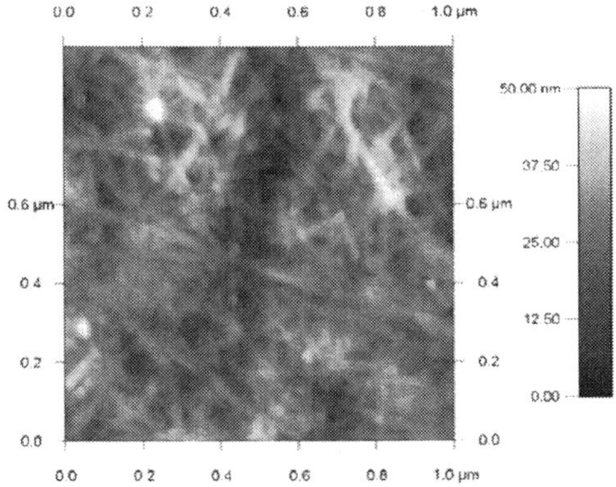

Figure 10. AFM images of five layers of CNT inkjet printed on silicon oxide substrate.

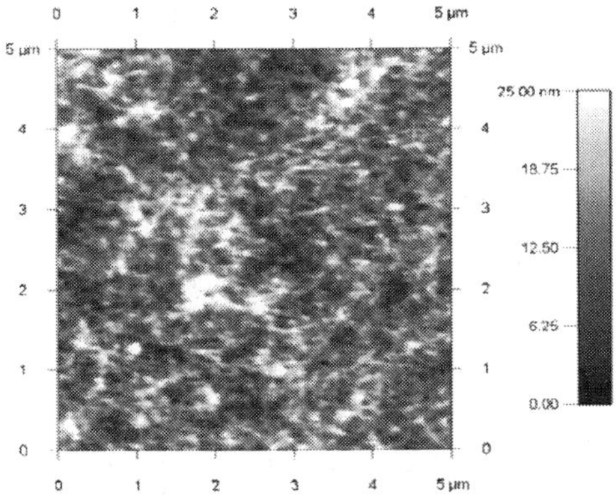

Figure 11. AFM images of five layers of CNT inkjet printed on Si substrate after 72 hours of soaking in DI water.

3.2.3. The resetting process of carbon nanotubes-based scale sensor

Figure 12 shows the ability of the resetting film (made of CNT inkjet printing) to reset the resistance values. The resetting process is the ability of the film to get back to its primitive condition after soaking it in DI water. The electrical resistance of the CNT 5 passes printed film is 1.0E+03 Ω; by dipping the film for one hour in DI water, the electrical resistance decreased to 5.0E+02 Ω. This treatment is needed to remove the left over ink from the surface as mentioned earlier. This value is the resistance of the optimized original sample and is thus used as the reference for the resetting process. Adding one drop of brine solution increases the electrical resistance up to 4.2E+03 Ω and decreases the sweeping voltage to 3.6E +03 Ω. By immersing the film for one hour in DI water, for the resetting process, we are able

http://dx.doi.org/10.5772/64196

to obtain an electrical resistance very close to the electrical resistance of the original optimized sample.

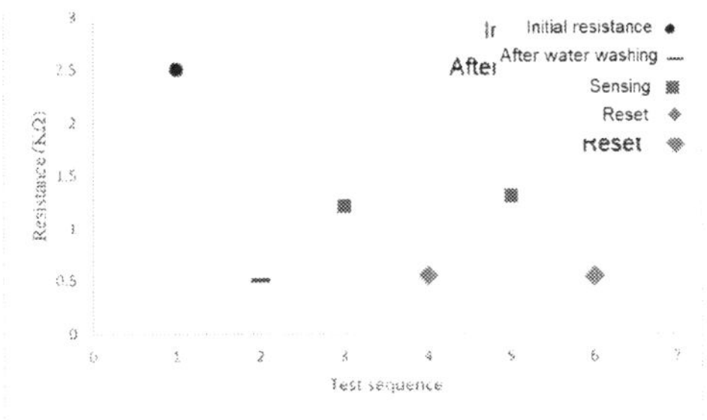

Figure 12. Ability to reset the resistance values for five layers of CNTs printed on silicon oxide substrate.

4. Conclusion

In summary, this chapter discussed the chemical sensing properties of CNTs obtained through smear casting and CNT inkjet printing techniques have been studied and compared. It was shown the ability of these films to be used for sensing salts, which are the precursors for the scale deposition. This was assessed through the modulation of the change in electrical conductance in the presence of various media. The SEM images showed that both CNT smear casted and the CNT inkjet printing were dispersed very well among the matrix and made a network structure. The results of these studies indicated that: (a) in case of casting, adding brine solution on the surface of the CNT casted films increased the resistance from 4.0E+3 kΩ to almost 9.5E+3 kΩ; (b) for the inkjet printing, five printing layers of CNT films was critical to obtain a consistent electrical resistance. Moreover, washing the original CNT inkjet printing sample with water was a precondition step to stabilize the CNT inkjet printing and ultimately get consistent electrical resistance. The resistance of the CNT inkjet printing increased from 0.50 kΩ to 1.50 kΩ once the brine solution was added to the surface of the sample. The CNT inkjet printing sample was found to be stable even after more than 48 hours of soaking in water.

Acknowledgements

The authors would like to thank the entire team of the Masdar Institute Cleanroom and microscopy suite for their contribution and support toward this work. In addition, we would like to acknowledge the help of Dr. Florent Favaux, and Khalid Marbou for SEM and AFM images. The authors would like also to appreciate Petroleum Institute for their partial funding of this research on behalf of the Abu Dhabi National Oil Company (ADNOC), Abu Dhabi, UAE.

Author details

Hammad Younes[1*], Amal Al Ghaferi[1] and Irfan Saadat[2]

*Address all correspondence to: hyounes@masdar.ac.ae; hasy193@yahoo.com

1 Department of Mechanical and Material Engineering, Masdar Institute of Science and Technology, Abu Dhabi, United Arab Emirates

2 Department of Electrical Engineering and Computer Science, Masdar Institute of Science and Technology, Abu Dhabi, United Arab Emirates

References

[1] Younes H, Christensen G, Liu M, Hong H, Yang Q, Lin Z. Alignment of carbon nanofibers in water and epoxy by external magnetic field. Journal of Nanofluids. 2014;3(1):33–7.

[2] Younes H, Christensen G, Li D, Hong H, Ghaferi AA. Thermal conductivity of nanofluids: review. Journal of Nanofluids. 2015;4(2):107–32.

[3] Christensen G, Younes H, Hong H, Peterson G. Alignment of carbon nanotubes comprising magnetically sensitive metal oxides by nonionic chemical surfactants. Journal of Nanofluids. 2013;2(1):25–8.

[4] Nguyen CV, Ye Q, Meyyappan M. Carbon nanotube tips for scanning probe microscopy: fabrication and high aspect ratio nanometrology. Measurement Science and Technology. 2005;16(11):2138–46.

[5] Abueidda DW, Dalaq AS, Al-Rub RKA, Younes HA. Finite element predictions of effective multifunctional properties of interpenetrating phase composites with novel triply periodic solid shell architectured reinforcements. International Journal of Mechanical Sciences. 2015;92:80–9.

[6] Neupane S, Khatiwada S, Jaye C, Fischer DA, Younes H, Hong H, et al. Single-walled carbon nanotubes coated by Fe_2O_3 nanoparticles with enhanced magnetic properties. ECS Journal of Solid State Science and Technology. 2014;3(8):M39–M44.

[7] Zhu Z, Hu Y, Jiang H, Li C. A three-dimensional ordered mesoporous carbon/carbon nanotubes nanocomposites for supercapacitors. Journal of Power Sources. 2014;246:402–8.

[8] Kalakonda P, Georgiev GY, Cabrera Y, Judith R, Iannacchione GS, Cebe P, editors. Thermal and electrical transport properties of sheared and un-sheared thin-film

polymer/CNTs nanocomposites. In MRS Proceedings 1660, mrsf13-1660. Cambridge University Press.

[9] Younes H. Carbon Nanomaterials and Metal Oxide Nanoparticles and Their Applications Toward Grease and Nanofluids. PhD Dissertation. 2014.

[10] Hong H, Thomas D, Waynick A, Yu W, Smith P, Roy W. Carbon nanotube grease with enhanced thermal and electrical conductivities. Journal of Nanoparticle Research. 2009;12(2):529–35.

[11] Hongtao L, Hongmin J, Haiping H, Younes H. Tribological properties of carbon nanotube grease. Industrial Lubrication and Tribology. 2014;66(5):579–83.

[12] Al-Saleh MH, Sundararaj U. Electromagnetic interference shielding mechanisms of CNT/polymer composites. Carbon. 2009;47(7):1738–46.

[13] Liu Z, Bai G, Huang Y, Ma Y, Du F, Li F, et al. Reflection and absorption contributions to the electromagnetic interference shielding of single-walled carbon nanotube/polyurethane composites. Carbon. 2007;45(4):821–7.

[14] Thomassin J-M, Vuluga D, Alexandre M, Jérôme C, Molenberg I, Huynen I, et al. A convenient route for the dispersion of carbon nanotubes in polymers: Application to the preparation of electromagnetic interference (EMI) absorbers. Polymer. 2012;53(1):169–74.

[15] Basuli U, Chattopadhyay S, Nah C, Chaki TK. Electrical properties and electromagnetic interference shielding effectiveness of multiwalled carbon nanotubes-reinforced EMA nanocomposites. Polymer Composites. 2012;33(6):897–903.

[16] Gupta TK, Singh BP, Mathur RB, Dhakate SR. Multi-walled carbon nanotube–graphene–polyaniline multiphase nanocomposite with superior electromagnetic shielding effectiveness. Nanoscale. 2014;6(2):842–51.

[17] Md. Mahfuzur Rahmana HY, George Ni, TieJun Zhang, Amal Al Ghaferi. Synthesis and optical characterization of carbon nanotube arrays. Materials Research Bulletin. 2016;77:243–52.

[18] Sinha N, Ma J, Yeow JT. Carbon nanotube-based sensors. Journal of Nanoscience and Nanotechnology. 2006;6(3):573–90.

[19] Wang J, Musameh M. Carbon nanotube screen-printed electrochemical sensors. Analyst. 2004;129(1):1–2.

[20] Han J-W, Kim B, Li J, Meyyappan M. A carbon nanotube based ammonia sensor on cellulose paper. RSC Advances. 2014;4(2):549–53.

[21] Gautam M, Jayatissa AH. Gas sensing properties of graphene synthesized by chemical vapor deposition. Materials Science and Engineering: C. 2011;31(7):1405–11.

[22] Gautam M, Jayatissa AH. Ammonia gas sensing behavior of graphene surface decorated with gold nanoparticles. Solid-State Electronics. 2012;78:159–65.

[23] Peng S, O'Keeffe J, Wei C, Cho K, Kong J, Chen R, et al., editors. Carbon nanotube chemical and mechanical sensors. In Proceedings of the 3rd International Workshop on Structural Health Monitoring 502–506. CRC Press, Stanford, CA, September 12–14, 2001.

[24] Crabtree M, Eslinger D, Fletcher P, Miller M, Johnson A, King G. Fighting scale—removal and prevention. Oilfield Review. 1999;11(3):30–45.

[25] Sofela SO, Younes H, Jelbuldina M, Saadat I, Al Ghaferi A, editors. Carbon nanomaterials based TSVs for dual sensing and vertical interconnect application. In Interconnect Technology Conference and 2015 IEEE Materials for Advanced Metallization Conference (IITC/MAM), 2015 IEEE International 289–292. IEEE France.

[26] Emmons, D.H., Graham, G.C., Holt, S.P., and Jordan, M.M. "On-site, near-real-time monitoring of scale deposition", SPE Annual Technical Conference and Exhibition, Houston, Oct. 1999, SPE 56776.

[27] Jordan MM, editor. Deployment of real-time scale deposition monitoring equipment to optimize chemical treatment for scale control during stimulation dlowback. In: SPE International Oilfield Scale Conference. Society of Petroleum Engineers, SPE Projects, Facilities & Construction, 6(1) 2008.

[28] Al-Matar H, Al-Ashhab JK, Mokhtar M, Ridzauddin S, editors. Techniques used to monitor and remove strontium sulfate scale in UZ producing wells. In: Abu Dhabi International Petroleum Exhibition and Conference. Society of Petroleum Engineers, Abu Dhabi International Petroleum Exhibition and Conference, 5-8 November, Abu Dhabi, UAE, ISBN 978-1-55563-173-4.

[29] Srivastava RK, Srivastava S, Narayanan TN, Mahlotra BD, Vajtai R, Ajayan PM, et al. Functionalized multilayered graphene platform for urea sensor. ACS Nano. 2011;6(1): 168–75.

[30] Borini S, White R, Wei D, Astley M, Haque S, Spigone E, et al. Ultrafast graphene oxide humidity sensors. ACS Nano. 2013;7(12):11166–73.

[31] Lu G, Ocola LE, Chen J. Gas detection using low-temperature reduced graphene oxide sheets. Applied Physics Letters. 2009;94(8):083111.

[32] Gonçalves V, Brandão L, Mendes A. Development of porous polymer pressure sensors incorporating graphene platelets. Polymer Testing. 2014;37:129–37.

Chapter 11

Corrosion Resistance and Tribological Properties of Epoxy Coatings Reinforced with Well-Dispersed Graphene

Liu Shuan

Additional information is available at the end of the chapter

http://dx.doi.org/10.5772/64097

Abstract

Seawater environment is the most harsh corrosion media, thus the deterioration of metal materials in marine environment is particularly serious. Organic coatings are good options for metal protection. Regarding the large impact of metal corrosion on marine engineering and the important significance of metal anticorrosion, this chapter deals with the research and development of the protective mechanism, wear resistance, and antifriction properties of the graphene-based epoxy coating used in marine environment. A highly efficient physical graphene dispersant was obtained by organic synthesis, and the graphene powder was dispersed uniformly in organic solvent and then added into epoxy resin. Moreover, the structure-activity relationship of the graphene-based epoxy coating was systematically investigated, taking into account its different amounts and dispersing characteristics in seawater. Consequently, the corrosion behavior and protection mechanism of the graphene-based heavy anticorrosion coating are evaluated and clarified. The obtained results are of fundamental importance for the field.

Keywords: grapheme, corrosion resistance, tribology, epoxy coating, dispersion

1. Introduction

The inhabitants of modernized nations live in metal-based societies. Almost all common metals or steels tend to react with their environments to some extents and at different corrosion rates. Corrosion is defined as the deterioration of a material, usually a metal, that requires the presence of an anode, a cathode, an electrolyte, and an electrical circuit [1]. Corrosion usually occurs as a natural phenomenon.

The sea occupies about three-quarters of the Earth and has several kinds of chemical ions. Seawater is a powerful natural corrosion environmental, which contains almost all of the elements present on the Earth. Some of these chemicals or ions (e.g., Cl^-, SO_4^{2-}) are corrosive towards metals. Corrosion damages considerably the marine steel infrastructures, such as ships, offshore oil installations, submarine pipelines, and port terminals, which not only causes serious economic losses but also threats the safety of human life and ecology [2, 3]. It has been estimated that the metal corrosion losses in China totaled one trillion and eight hundred billion in 2014, which accounted for one-third of the average gross national product (GNP). Marine corrosion accounts for the majority of metal losses. Thus, an effective protection technology is necessary to inhibit corrosion. Up to now, several methods have been used to protect metals against corrosion such as cathodic protection, protective coating, corrosion inhibitors, and addition of alloy elements. Among those procedures, coating is the most effective and economical way to inhibit metal corrosion, as it can block the contact of the corrosive medium with the metal [4].

1.1. Anticorrosion and failure mechanism of organic coating

Organic coatings can be found everywhere in industrialized nations, in order to protect metal structures and metal equipment. They can protect metals from corrosion in two ways. First, the coating film acts as a physical barrier to separate the metal substrate from the corrosive environment. Second, the organic coating can contain corrosion inhibitors, which decrease the corrosion rate of the metal.

However, organic coatings are not perfect barriers as they can be penetrated by corrosive media, such as oxygen, water molecule, and ions. When this happens, metal corrosion can occur underneath the organic coating/metal interface. Thus, the corrosive media accumulated beneath the organic coating not only leads to metal corrosion but also decreases the adhesion of the coating to the metal. Moreover, the corrosion products formed on the metal surface can also destroy the coating/metal system. Generally, there are three methods often used to improve the protective performance of the organic coatings. First, fillers or pigments, such as graphene, nanoclay, ZnO, and MoS_2, are added into the coating to extend the diffusion path of corrosive media through the coating. Second, corrosion inhibitors are added to inhibit metal corrosion reaction underneath the coating. Third, high density, high cross-linked, thicker, and self-healing coating are needed to reduce the coating permeability [5].

1.2. Graphene based epoxy coating used for metal protection

Graphene (G) is a two-dimensional material with a single layer of around 0.335 nm and a diameter ranging from a few to several hundred microns. It has attracted increasing attention from academic and industrial fields due to their unique nanostructure, excellent physical properties, large specific surface area, super hydrophobic properties, and good compatibility with polymer matrixes [6, 7]. It has been reported that G can exhibit extraordinary corrosion resistance and self-lubricant characteristics due to its unique flexible graphitic layers, super-hydrophobic characteristics, extremely high strength, and easy shear capability on its densely packed and atomically smooth surface [8]. However, G nanosheets are prone to aggregate due to their strong van der Waals forces and high specific surface areas, which is difficult in achieving an homogenous dispersion of G in organic coatings. Many works on chemical modification approach have been reported to improve dispersion of G in composite coatings [9–11]. For example, Chang et al. [12] reported on the preparation of polyimide grafted G via a thermal imidization reaction. The well-dispersed G/polyimide coatings provided advanced corrosion protection of cold-rolled steel, as compared to neat polyimide coating. Compared to chemical modification, simply dispersing G into a polymer matrix can be particularly advantageous due to less damage made to G, and the high efficiency and ease of the preparation method.

In this chapter, the fabrication of composite coatings (with epoxy coating as the organic matrix and uniform dispersed G as anticorrosive barrier) is reported. The corrosion resistance and tribological properties of epoxy coatings reinforced with well-dispersed G are evaluated in details.

2. Corrosion resistance and tribological properties of G/epoxy composite coating

2.1. G powder dispersion performance in organic solvent

G was supplied by Ningbo Morsh Technology Company from China. Four different physical G dispersants, namely dispersant-1 (it has a basic triphenylmethane structure), dispersant-2 (it is a sodium polyacrylate), dispersant-3 (it is a sodium hexametaphosphate), dispersant-4 (it has a carboxylated oligoaniline structure), were synthesized and used to disperse G. **Figure 1** depicts the G powder dispersed in ethanol solution by the different G dispersants after 30 days. Without any dispersant, the G powder was aggregated and quickly precipitated at the bottom of the container (**Figure 1a**). Dispersants-1 can be adsorbed on the surface of G, however, the Dispersants-1/G system precipitated after a few hours (**Figure 1b**); for Dispersants-2 and Dispersants-3, G powder was swelled and can be dispersed, but G precipitated after 12 h (**Figure 1c** and **d**). Dispersant-4 was able to stably disperse G in ethanol after even 30 days (**Figure 1e**), only a few G slurry can be found at bottom of glass bottle. Interestingly, when the dispersed G was dried by spraying, it still could be re-dispersed in several solvents, such as water, tetrahydrofuran (THF), and ethanol (**Figure 2**). **Figure 3** shows the SEM images of dispersed G in ethanol solution. It can be seen that the layered G was completely dispersed at length of ~5 μm, and no obvious aggregates existed. **Figure 4** depicts an atomic force microscopy (AFM) image of dispersed G deposited on a Si-coated substrate. On the topography image (**Figure 4a**), the G layer can be seen as slightly brighter than the Si-coated substrate. From the line scan shown in **Figure 4b**, it is very easy to estimate the thickness of G layer (about 3–6 nm), indicating that the aggregated G was exfoliated to ultra-thin layers.

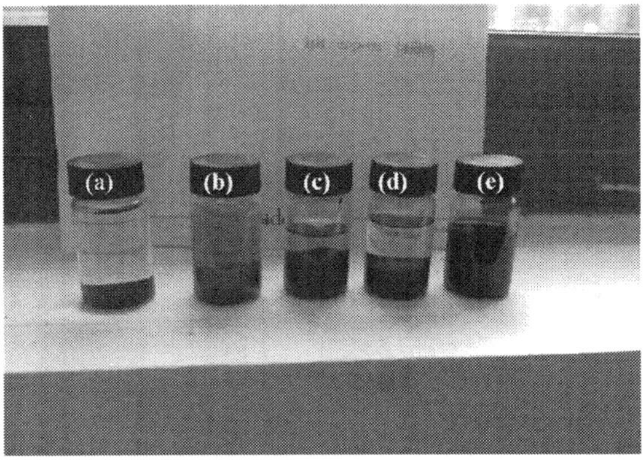

Figure 1. G powder dispersed in ethanol solution using different G dispersants after 30 days; the mass ratio of dispersant to G powder was 1:5. (a) G+ethanol; (b) G+dispersant-1+ethanol; (c) G+dispersant-2+ethanol; (d) G+dispersant-3+ethanol; (e) G+dispersant-4+ethanol.

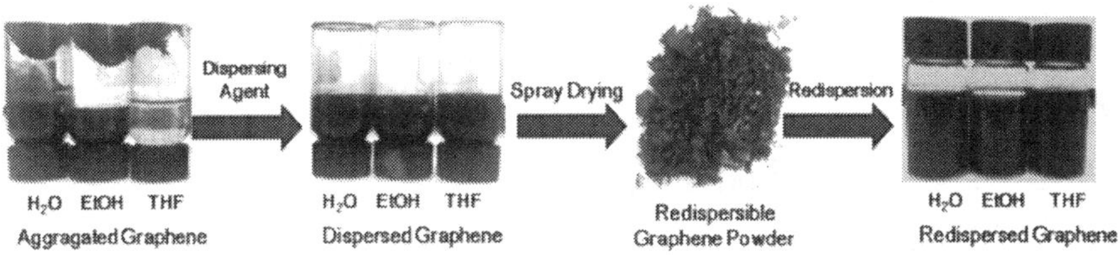

Figure 2. Schematic diagram of G dispersion and re-dispersion in different solvents.

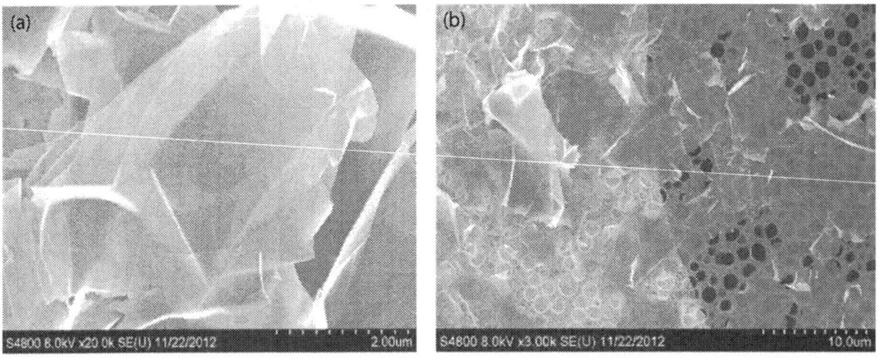

Figure 3. SEM images of dispersed G at low magnification (a) and high magnification (b).

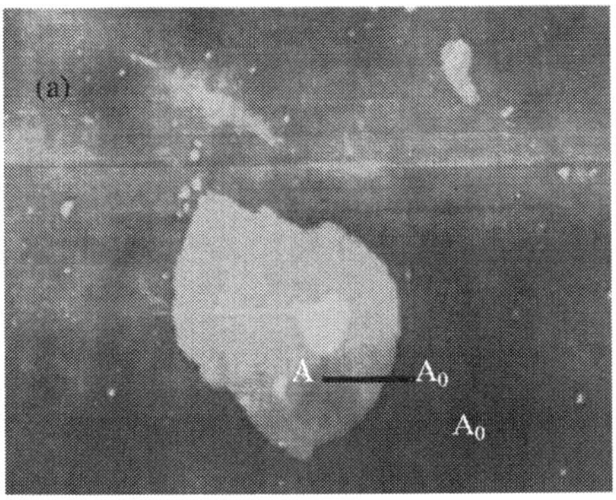

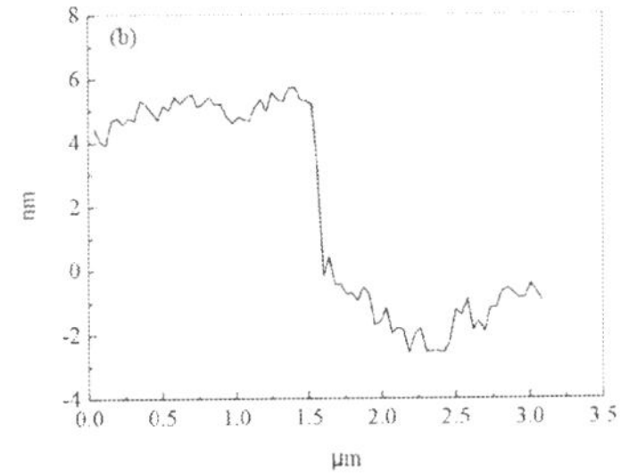

Figure 4. AFM images (a) of a thin G film deposited on a Si-coated substrate and height profile (b) along the black line (from A to A_0).

2.2. Preparation of a G/epoxy composite coating

Figure 5 depicts the preparation process of G-based epoxy coatings. The well-dispersed G was mixed in the epoxy resin and cured by a polyamide agent (650 polyamide curing agent, light yellow, Amine value: 420 mgKOH/g, viscosity: 18,000 mPa s/40°C). **Figure 6a** and **b** shows typical SEM micrographs of fracture surfaces (cross section) of an epoxy coating containing 0.5 wt% G. It is well known that the fracture surface of pure epoxy thermoset exhibits oriented "bamboo"-like fracture patterns starting from the cracks. In contrast, this G based epoxy coating showed less brittle-like fracture morphology. The existence of highly dispersed G effectively disturbs the crack propagation due to its two dimensional nanostructure. **Figure 7** depicts TEM images of G (0.5 wt%) in epoxy coatings. It is shown that G is randomly dispersed in the coatings, and the good dispersion of G can extend the diffusion path of corrosive media in the epoxy coating.

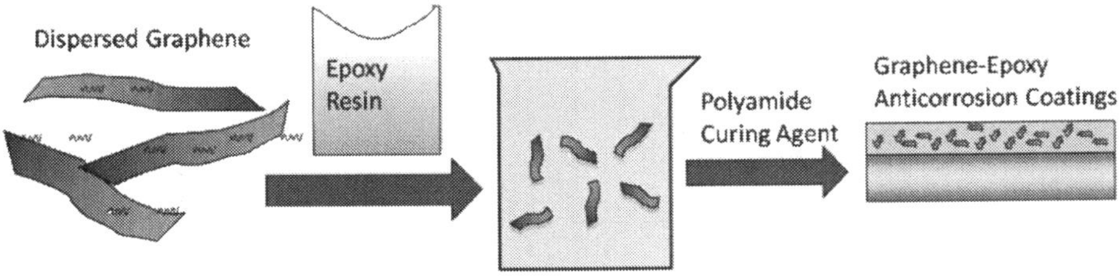

Figure 5. Preparation process of G/epoxy composite coatings.

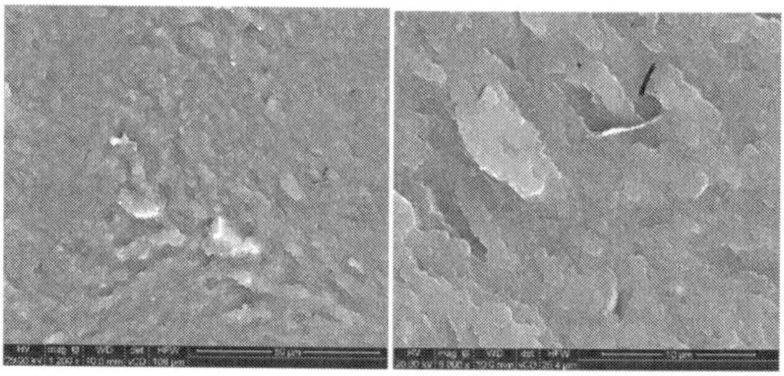

Figure 6. SEM images of cross sections of a G/epoxy composite coating (0.5 wt% G content).

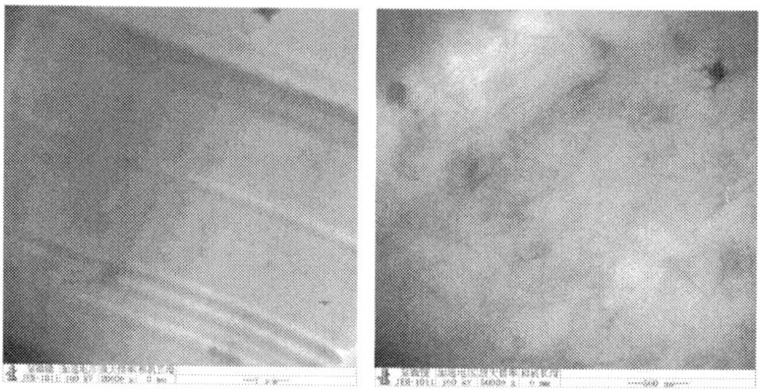

Figure 7. TEM images of a G/epoxy composite coating (0.5 wt% G content).

2.3. Corrosion resistance and mechanism of G/epoxy composite coating

In order to evaluate the improvement obtained with the G/epoxy composite coating, the pure epoxy coating, the D4/epoxy coating, and the G/epoxy coating were prepared in a similar way to the described above for the G-D4/epoxy coating (D4 means dispersant 4, it has a carboxylated oligoaniline structure). Electrochemical measurements were performed on a CHI-660E electrochemical workstation in 3.5% NaCl solution at room temperature. A three-electrode setup was used for EIS test, a saturated calomel electrode (SCE) was used as the reference electrode, a platinum plate of 2.5 cm^2 as the counter electrode, and coating/Q235 steels as the working electrode. Before electrochemical impedance spectroscopy (EIS) measurement, the coating/Q235 steel specimen was initially kept at an open circuit potential (OCP) for 1 h until a stable state was attained. For EIS, the test frequency range was 10^5 to 10^{-2} Hz and the amplitude of the sinusoidal voltage signal was 20 mV. ZsimpWin 3.21 software was used to fit the EIS spectra.

Figure 8 shows the open circuit potential (E_{OCP}) variation of pure epoxy, G/epoxy, D4/epoxy, and G-D4/epoxy as a function of immersion time in 3.5% NaCl solution. The E_{OCP} of all coatings decreased to different degrees after the fluctuation at initial time and then remained basically unchanged after 16 days. Compared with pure epoxy, D4/epoxy and G-D4/epoxy exhibit a larger E_{OCP}, whereas G/epoxy displays a E_{OCP} similar to that of the pure epoxy. The positive shift of E_{OCP} for 0.30 V for G-D4/epoxy and 0.10 V for D4/epoxy after 35-day immersion shows that the G-D4 hybrid and D4 have obvious corrosion inhibition properties, and a similar phenomenon was reported by other researchers [13]. Furthermore, the G-D4 hybrid exhibits better corrosion inhibition than D4, while commercial G without D4 functionalization has almost no corrosion inhibition.

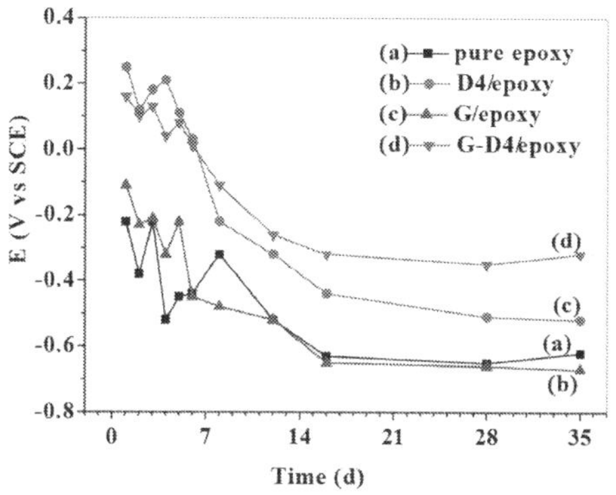

Figure 8. Time dependence of the open circuit potential (E_{OCP}) of pure epoxy (a), G/epoxy (b), D4/epoxy (c) and G-D4/epoxy (d) coated Q235 steel electrodes immersed in 3.5% NaCl solution.

EIS was further employed to investigate the corrosion mechanism and evaluate the degradation processes of the G-D4/epoxy composites, as shown in **Figure 9**. The Bode phase plots show one time constant at the initial stage of immersion (stage I), and then two time constants appeared due to the penetration of corrosive media during the immersion process (stage II). Usually, the time constant at the high frequency range can be attributed to the coating layer, while the one time constant at lower frequencies corresponds to a corrosion process taking place at the metal/coating interface. The pure epoxy, G/epoxy, and D4/epoxy coatings exhibit two time constants in short term immersion (2 days, 1 day, and 4 days, respectively), while the G-D4/epoxy coating shows one time constant in a high frequency range for 28-day immersion. The impedance modulus at low frequency ($|Z|_{0.01Hz}$) can represent the corrosion protection of coating/metal system, which is in inverse proportion to corrosion rate. The $|Z|_{0.01Hz}$ of the G-D4/epoxy-coated Q235 electrode was 1.6×10^9 Ω cm^2 after 16-day immersion (**Figure 9D**), which was far larger than that of pure epoxy (2.2×10^7 Ω cm^2), G/epoxy (3.1×10^6 Ω cm^2), and D4/epoxy (5.5×10^7 Ω cm^2)-coated electrodes after 12-day immersion (**Figure 9A–C**). These results indicate that the G-D4/epoxy coating provides better corrosion protection for the Q235 steel than other coatings. The increase in impedance for G-D4/epoxy coating can be attributed to the hydrophobicity and barrier effects of the well dispersed graphene.

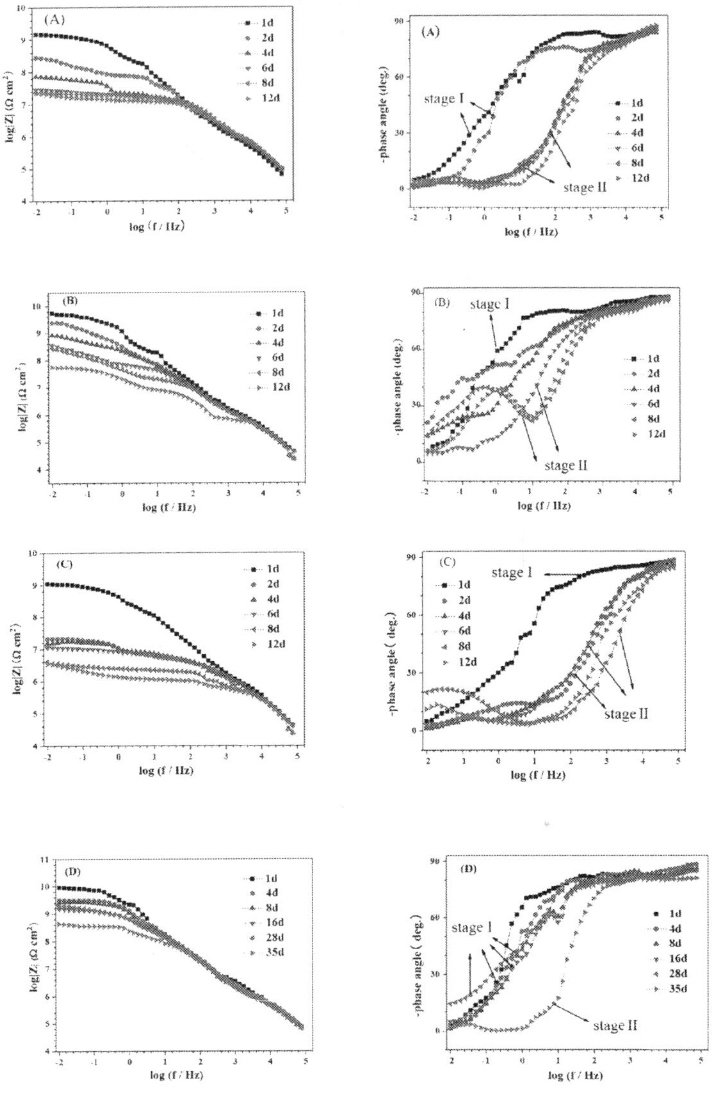

Figure 9. Bode plots of pure epoxy (A), D4/epoxy (B), G/epoxy (C), and G-D4/epoxy (D) coated Q235 steel electrodes immersed in 3.5% NaCl solution after different times.

Water uptake of the coating was calculated based on the changes of the coating capacitance due to the variation of the dielectric constant. The water diffusion coefficient *D* could be calculated by means of the simplified Fick's second law of diffusion [14] (see Eq. (**1**)):

$$\frac{\lg Q_c - \lg Q_0}{\lg Q_\infty - \lg Q_0} = \frac{2}{L}\sqrt{\frac{D}{\pi}}\sqrt{t} \tag{1}$$

where Q_0, Q_c, and Q_∞ are the coating capacitances at the beginning of the immersion time t_0, the time t_c, and the time in saturated water absorption state t_∞, respectively. D is the diffusion coefficient, L is the coating thickness. The coating capacitances obtained from the fitted corrosion parameters, were fitted with equivalent circuits (**Figure 10**) by ZsimpWin 3.21 software. D can be calculated from the $\lg Q_c - t^{1/2}$ curves (**Figure 11**), yielding the water diffusion coefficients of these 4 coating (pure epoxy, G/epoxy, D4/epoxy, and G-D4/epoxy), calculated to be 2.25×10^{-12}, 23.6×10^{-12}, 17.1×10^{-12} s and 0.619×10^{-12} cm²/s, respectively (**Figure 12**). The significant decrease of water diffusion coefficient for G-D4/epoxy coating confirmed that the water penetration into the epoxy coating was impeded by the highly dispersed G, while the G/epoxy coatings without D4 exhibited much higher water diffusion coefficient due to the aggregation of G.

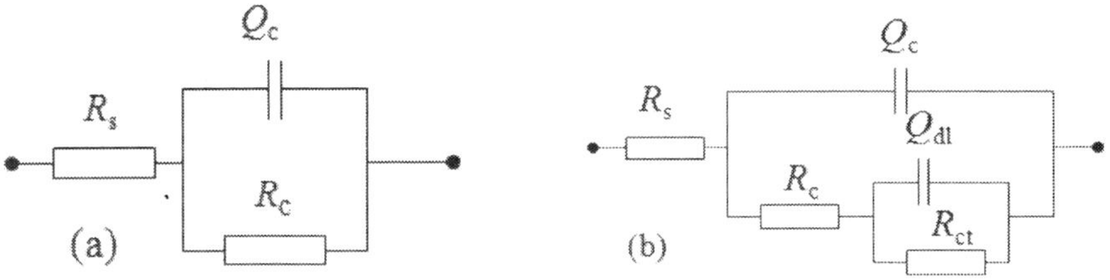

Figure 10. The equivalent circuit (a) for stage I and (b) for stage II used to fit the EIS data.

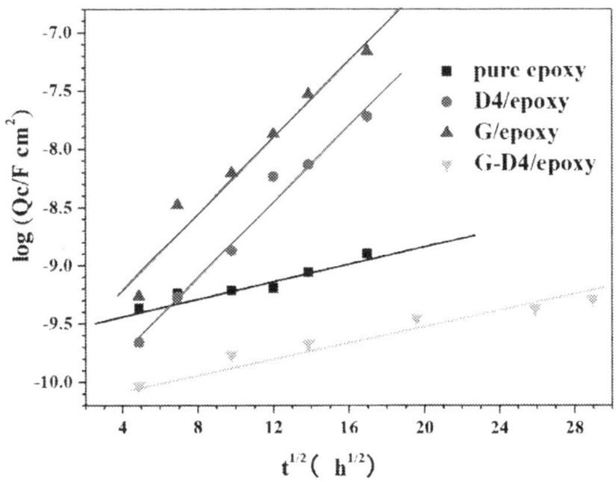

Figure 11. $\lg Q_c - t^{1/2}$ curves of different composite coatings immersed in 3.5% NaCl solution.

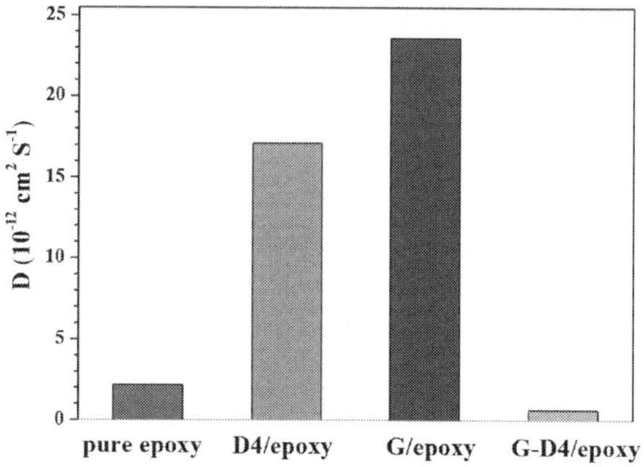

Figure 12. The water diffusion coefficient D of different composite coatings immersed in 3.5% NaCl solution.

2.4. Tribological properties of graphene/epoxy composite coating

The coefficient of friction (COF) was conducted on UMT-3 tribometer (UMT-3, Bruker-CETR, USA) with a ball-on-plate configuration using the following conditions: load of 3N, sliding velocity of 1 Hz, friction duration of 20 min, and wear track length of 5 mm. All the experiments were performed under dry condition and seawater lubricated condition. **Figure 13** shows the friction coefficient of the G/epoxy coatings as function of test time under both of dry sliding and seawater lubrication, respectively. It was clear that the COF of G/epoxy coatings initially increases rapidly and then reaches stability with time in dry conditions, but in seawater environment, the COF of G/epoxy coatings initially decrease and then reach a stable state. The COF values of neat epoxy coating, 0.25%G-epoxy coating, and 0.5%G-epoxy coating were 0.49,

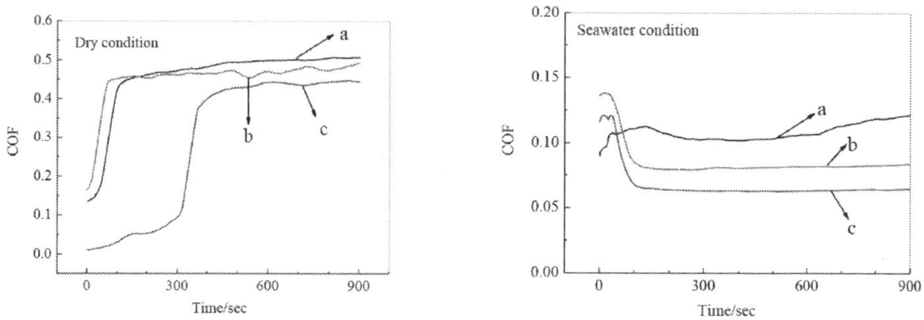

Figure 13. COF of three G-epoxy coating in dry and seawater friction conditions. (a) Neat epoxy coating; (b) 0.25%G-epoxy coating; (c) 0.5%G-epoxy coating.

0.45, and 0.42 in dry conditions, respectively. The COF values of neat epoxy coating, 0.25%G-epoxy coating, and 0.5%G-epoxy coating were 0.12, 0.08, and 0.06 in seawater, respectively. These results suggest that G can decrease the COF values of epoxy coatings in both dry and seawater conditions.

The wear traces of the different coatings were evaluated by using a surface profiler (Alpha-Step IQ, KLA-Tencor, USA). The wear resistance of the coatings is represented by wear trace cross-sectional area on account of the uniform wear trace length. **Figure 14** shows the wear trace areas of G/epoxy coatings with different mass ratio of G measured under dry sliding and seawater lubrication. It can be seen that the wear trace areas of composite coatings decreased with the increase of G content in G/epoxy coatings. In addition, the COF and wear traces area, under seawater lubrication, were much lower than those under dry sliding, during the whole process. Three main factors can cause these phenomena: First, the lubricating effect of seawater can effectively reduce the direct contact of the composite coatings with the counterpart surface. Second, the seawater medium can act as coolant which significantly reduces the friction heat, avoiding thermal softening, and resulting in lower COF and wear traces area. Finally, the wear debris can be continuously removed by seawater, leading to smooth and clean worn surface.

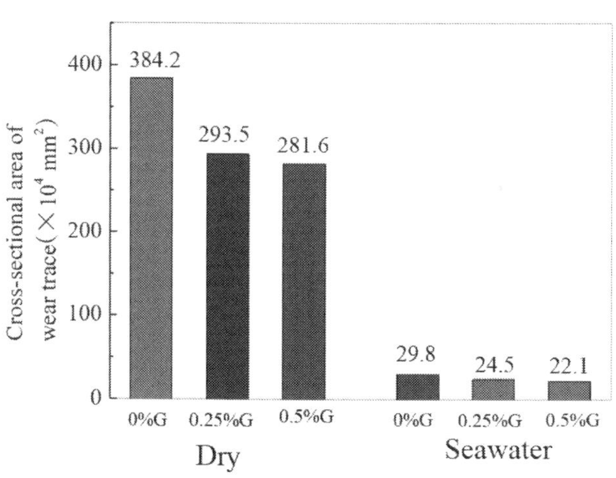

Figure 14. Cross-sectional area of wear trace of three G-epoxy coatings in dry and seawater friction conditions.

3. Conclusions

1. Graphene (G) can be uniformly dispersed in several solvents.

2. The well-dispersed G can block the coating pores and decrease the electrolyte diffusion towards the epoxy coating, the water diffusion coefficients of epoxy coating decrease with addition of 0.5 wt% G.

3. G increases the corrosion resistance of epoxy coatings.

4. Friction coefficient and wear rate of the epoxy coating decrease with the addition of G; the friction coefficient and wear rate of G/epoxy composite coating in seawater was lower than in dry conditions.

Acknowledgments

The authors wish to acknowledge the financial support of the National Science Foundation of China (No. 41506098), China Postdoctoral Science Foundation (No. 2015M580528), Zhejiang Province Preferential Postdoctoral Science Foundation (No. BSW1502160) and Open Fund Project of Key Laboratory of Marine Materials and Related Technologies (LMMT-KFKT-2014-008) in Chinese Academy of Sciences.

Author details

Liu Shuan

Address all correspondence to: liushuan@nimte.ac.cn

Ningbo Institute of Materials Technologies and Engineering, Chinese Academy of Sciences, Ningbo, China

References

[1] E McCafferty. Introduction to Corrosion Science. Springer, Alexandria Va, USA ,2009.

[2] S Liu, H Y Sun, N Zhang, et al. The corrosion performance of galvanized steel in closed rusty seawater. International Journal of Corrosion, 2013 (2013) 1–9.

[3] S Liu, H Y Sun, L J Sun, et al. Effects of pH and Cl^- concentration on corrosion behavior of the galvanized steel in simulated rust layer solution. Corrosion Science, 65 (2012) 520–527

[4] L Gu, S Liu, H C Zhao, et al. Anticorrosive oligoaniline-containing electroactive siliceous hybrid materials. RSC Advance, 5 (2015) 56011–56019.

[5] B Chen, J Ma, L Gu, et al. Anticorrosive properties of oligoaniline containing photo-cured coatings. International Journal of Electrochemical Science, 10 (2015) 9154–9166.

[6] L Gu, S Liu, H C Zhao, et al. Facile Preparation of water-dispersible graphene sheets stabilized by carboxylated oligoanilines and their anticorrosion coatings. ACS Applied Materials and Interfaces, 7 (2015) 17641–17648.

[7] S Liu, L Gu, H C Zhao, et al. Corrosion resistance of graphene-reinforced waterborne epoxy coatings. Journal of Materials Science & Technology, 2013, 32(5) : 425–431.

[8] C M Kumar, T V Venkatesha, R. Shabadi. Preparation and corrosion behavior of Ni and Ni–graphene composite coatings. Materials Research Bulletin, 48 (2013) 1477–1483.

[9] B Ramezanzadeh, E Ghasemi, M Mahdavian, E Changizi, M H Mohamadzadeh. Covalently-grafted graphene oxide nanosheets to improve barrier and corrosion protection properties of polyurethane coatings. Carbon, 93 (2015) 555–573.

[10] C H Chang, T C Huang, C W Peng, et al. Novel anticorrosion coatings prepared from polyaniline/graphene composites. Carbon, 50 (2015) 5044–5051.

[11] Q F Jing,W S Liu, Y Z Pan, et al. Chemical functionalization of graphene oxide for improving mechanical and thermal properties of polyurethane composites. Materials and Design 85 (2015) 808–814.

[12] K C Chang, C H Hsu, H I Lu, et al. Advanced anticorrosive coatings prepared from electroactive polyimide/graphene nanocomposites with synergistic effects of redox catalytic capability and gas barrier properties. Express Polymer Letters, 8 (2014) 243–255.

[13] H Wei, D Ding, S Wei, et al. Anticorrosive conductive polyurethane multiwalled carbon nanotube nanocomposites. Journal of Materials Chemistry A, 1 (2013) 10805–10813.

[14] M M Wind, H J W Lenderink. A capacitance study of pseudo-Fickian diffusion in glassy polymer coatings. Progress in Organic Coatings, 28 (1996) 239–250.

Chapter 12

Fully Configurable Electromagnetic Wave Absorbers by Using Carbon Nanostructures

Davide Micheli, Roberto Pastore,
Antonio Vricella and Mario Marchetti

Additional information is available at the end of the chapter

http://dx.doi.org/10.5772/64213

Abstract

The configurable electromagnetic wave absorber (CEMA) defines a new method for the full design of layered carbon-based nanocomposites able to quasi-perfectly reproduce any kind of EM reflection coefficient (RC) profile. The method involves three main factors: (a) nanofillers-like carbon nanotube (CNT), carbon nanofiber (CNF), graphene nanoplatelet (GNP), and polyaniline (PANI) in different concentration versus the matrix; (b) the dielectric parameters of the nanoreinforced materials in the microwave range 2–18 GHz; (c) a numerical technique based on particle swarm optimization (PSO) algorithm within the MATLAB code of the EM propagation engine. Output is the layering of the wave absorber, that is, number of layers and material/thickness of each layer and the reflection/transmission simulated profiles. The frequency selective behavior is due to the multilayered composition, thanks to the direct/reflected wave combination tuning at interfaces. The dielectric characterization of the employed nanocomposites is presented in details: these materials constitute the database for the optimization code running toward the multilayer optimal solution. A FEM analysis is further proposed to highlight the EM propagation within the material's bulk at different frequencies. The mathematical model of layered materials, the PSO objective function used for RC target fitting, and some results are reported in the text.

Keywords: carbon nanoparticles, particle swarm optimization, layered electromagnetic wave absorber, nanocomposite materials

1. Introduction

In the last 50 years, many research activities have been focused on materials and structures able to reduce the electromagnetic reflection coefficient (RC) in certain frequency ranges.

Applications are mostly in military technology, with radar absorbing materials (RAM) and radar absorbing structure (RAS) [1–6], and in electromagnetic compatibility (EMC) for EM shielding and EM interference (EMI) suppression purposes [7–9], as well as for antenna testing in anechoic chambers [10, 11]. Nowadays, the increase of wireless communication systems demands the use of specific technologies or technical solutions to reduce the mutual telecommunication (TLC) systems interference. In fact, in order to reduce the energy impact and the constraints imposed by cost savings, most different wireless systems are often co-located in the same place, that is, sharing the same basic infrastructures. The proximity of such TLC systems can give rise to mutual radio frequency (RF) interferences due to intermodulation products in the shared antenna systems and/or to low efficiency in filtering out the spurious RF components. In this complex context, the availability of a new kind of materials, able to be frequency selective, absorbing, and tunable, is particularly attractive to reduce mutual EMI issues in specific frequency ranges [12]. Another field where configurable EM wave absorbers appear useful is the metrological science and technology. For example, in remote sensing, operations are sometimes necessary to often verify the sensors capability to detect and recognize certain material signatures on the Earth surface [13, 14]. Such type of testing and better sensors and algorithms calibration could be easily tuned in situ by using standard metrological samples, able to reproduce the desired EM target signatures. Further applications of tunable EM wave absorbers are conceivable in the security field. Nowadays, some studies are focused on the detection of explosives using appropriate radar signatures [15, 16]: the possibility to design metrological samples able to reproduce some particular explosive matter radar signatures could represent an interesting aid for the safety upgrade in strategic/critical environment.

This chapter is composed of three main sections. The first describes the nanocomposite materials database obtained from the dielectric characterization, by means of vector network analyzer (VNA) measurements, the mathematical modeling of the layered absorber, and the objective function optimization, used by the PSO algorithm. The second reports the main results of simulations, related to several ideal RC targets addressed; the third provides a discussion of the results based upon a FEM numerical analysis by COMSOL Multiphysics code.

2. Materials and methods

2.1. Nanocomposite materials

Several microwave reflection profiles, even having complicated shapes, have been devised to design absorbers with fitting EM responses. The proposed strategy to achieve the target behavior is to combine different components from the microwave materials database in a multilayer assembly. Such a soft approach allows to exploit a variety of dielectric properties provided by the available constituents in the materials database, as well as the characteristics of the EM field propagation, through the physical discontinuities occurring at the several layers interfaces. Polymeric composites filled with carbon nanoparticles provide an ideal platform in this framework, thanks to the possibility to explore a wide range of EM trans-

Figure 1. Sample holder used to build ring-shaped samples of nanocomposite having dimensions of 3 mm and 7 mm for inner and outer diameter respectively.

port properties, from the insulating matrix to rather conductive composites, just by tuning the nano-filler concentration within the basic polymer. A materials database of 23 carbon micro-powder- and nano-powder-filled polymeric composites has been set up and exploited. Such materials were obtained using commercial low cost products, in a well-established time-saving manufacturing procedure, in view of possible forthcoming scaling-up developments. In particular, a low-viscosity epoxy resin (Prime™ 20 LV—Gurit) was employed as polymeric matrix, while industrial grade multi-walled carbon nanotubes (MWCNT, NC7000—Nanocyl), carbon nanofibers (CNF—Sigma–Aldrich), graphene nanoplatelets (GNP, C750—XG Science), and polyaniline emeraldine base powder (PANI—Sigma–Aldrich) were added to the matrix as reinforcing materials, in different concentrations (from 0 up to 3 wt.%). The composite samples were shaped by means of an ensemble of hollow cylindrical molds (**Figure 1**) with 3 mm of internal diameter and 7 mm of the external one, as required by the specific technique adopted for the microwave characterization. The latter is based on the transmission line by means of coaxial air-line 50 Ω probe and has been carried out by means of a vector network analyzer (VNA, Agilent PNA-L N5235), via coaxial airline method in the frequency range of 2–18 GHz. The output of the measurements is obtained by means of the samples microwave reflection/transmission parameters and the materials complex electrical permittivity; the other typical dielectric properties of interest (loss tangent, skin depth, intrinsic impedance, etc.) can be then easily retrieved. The magnetic permeability of all the materials has not be taken into account during the analysis, since it is always very close to 1 (as expected for non-magnetic materials), as confirmed by the experiments. In **Figure 1**, the sample holder used to build ring samples of different reinforced materials is shown. Each hole of the sample holder is able to host the poured composite after the mixing and sonication procedures; after the curing process, the sample holder is open and the samples can be easily extracted. In **Figure 2**, the 50 Ω coaxial airline used for the VNA scattering parameters [17–19] measurements and for the subsequent computation of the relative dielectric parameters is shown: the samples are inserted around the central conductor of the coaxial airline.

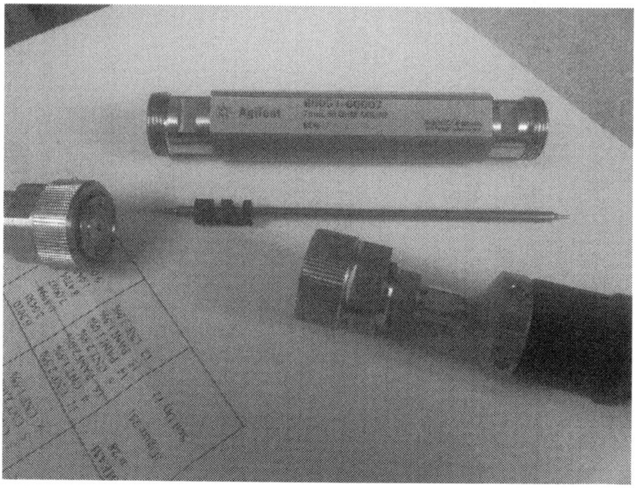

Figure 2. Coaxial airline 50 Ω probe used for the dielectric characterization of nanocomposite materials and for the measurement of EM reflection and transmission properties of layered materials.

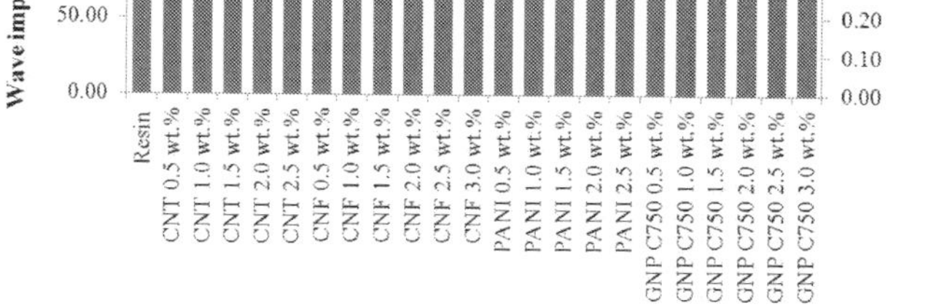

Figure 3. Loss tangent and intrinsic wave impedance of the tested nanocomposite materials in the database averaged over the frequency band 2–18 GHz.

Figure 3 shows the loss tangent and the intrinsic wave impedance of the nanocomposite materials present in the database, averaged in the frequency band 2–18 GHz. It can be seen that the CNT- and CNF-filled composites show higher EM losses and lower wave impedances, thus anticipating the employment of such materials in layers where a significant EM absorption capability is required; on the other hand, the behavior of PANI/GNP-filled materials is similar to that of the naked resin, thus suggesting the use of such materials in impedance matching layers. The behavior of the electric permittivity of CNT-based nanocomposites is reported in **Figures 4** and **5** (real and imaginary parts, respectively). In the plots, a remarkable variation of permittivity can be observed upon variation of the CNT weight concentration, according to the change of resistive losses due to the conductive filler inclusion: such trend is more evident at lower frequencies, due to skin-depth effects. In **Table 1**, the database of materials is listed by the code number, thereafter used. In **Table 1**, the dielectric complex permittivity averaged over the microwave range investigated is also reported to allow the readers a quick idea about the materials properties. A detailed description of the materials manufacturing procedure, as well as the full database microwave characterization plotting and analysis, is available in a previously published work [20].

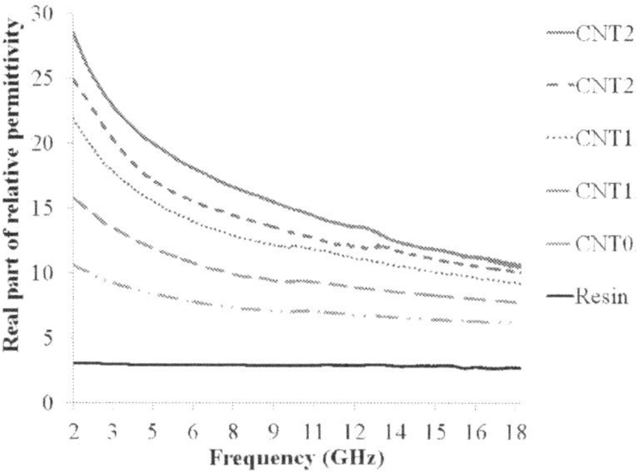

Figure 4. Real part of permittivity of CNT-based nanocomposites.

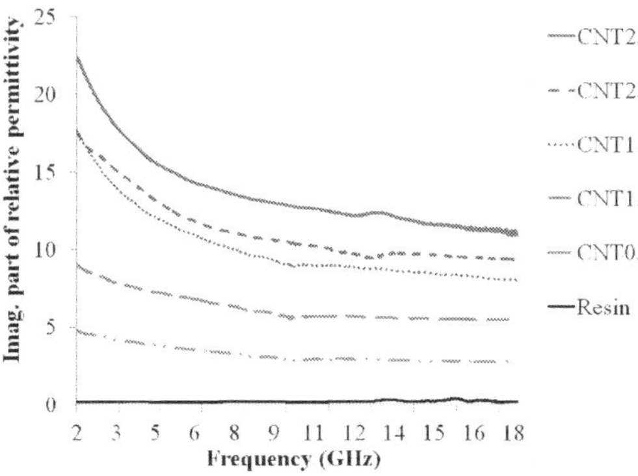

Figure 5. Imaginary part of permittivity of CNT-based nanocomposites.

Code	Material	Average complex permittivity
1	Resin PRIME TM	2.88–j0.19
2	CNT 0.5 wt%	7.36–j3.27
3	CNT 1.0 wt%	9.98–j6.27
4	CNT 1.5 wt%	12.76–j10.14
5	CNT 2.0 wt%	14.14–j11.21
6	CNT 2.5 wt%	15.87–j13.66
7	CNF 0.5 wt%	3.74–j0.38
8	CNF 1.0 wt%	4.39–j0.53
9	CNF 1.5 wt%	4.92–j0.69
10	CNF 2.0 wt%	5.97–j1.02
11	CNF 2.5 wt%	7.57–j1.74
12	CNF 3.0 wt%	9.74–j2.99
13	PANI 0.5 wt%	3.20–j0.21
14	PANI 1.0 wt%	3.16–j0.26
15	PANI 1.5 wt%	3.31–j0.28
16	PANI 2.0 wt%	3.09–j0.26
17	PANI 2.5 wt%	3.15–j0.28
18	GNP 0.5 wt%	2.90–j0.16
19	GNP 1.0 wt%	3.17–j0.21
20	GNP 1.5 wt%	3.20–j0.20
21	GNP 2.0 wt%	3.27–j0.20
22	GNP 2.5 wt%	3.36–j0.24
23	GNP 3.0 wt%	3.44–j0.27

Table 1. Microwave composite database: materials code, composition, and complex permittivity averaged in the frequency range 2–18 GHz.

2.2. Mathematical model and objective function optimization

The multilayered EM absorber design is based on the mathematical modeling of an EM wave propagating through a sequence of slabs made of different materials. Since the goal is to reproduce a given microwave reflection profile, it is worth to remind that the reflection coefficient (*RC*) at an air-matter interface can be evaluated by the relationship:

$$RC_i = \left| \frac{Z_i - Z_0}{Z_i + Z_0} \right| \tag{1}$$

which relates the free-space impedance ($Z_0 \approx 377\ \Omega$) to the input impedance (Z_i) seen at the impinged surface for a given ith value of frequency. Such quantity depends on the dielectric properties of the material; for a multilayer structure, it is affected by multiple reflections occurring at the material interfaces and can be obtained by iterating Eq. (1) layer by layer through the formalism:

$$Z_{iu} = \eta_{iu} \frac{Z_{iu-1}\cos(k_{iu}t_u) + j\eta_{iu}\sin(k_{iu}t_u)}{\eta_{iu}\cos(k_{iu}t_u) + jZ_{iu-1}\sin(k_{iu}t_u)} \tag{2}$$

$$\eta_{iu} = \sqrt{\frac{\mu_0}{\varepsilon_0}}\sqrt{\frac{\mu_{ru}}{\varepsilon_{ru}}} \tag{3}$$

$$k_{iu}^2 = (2\pi f)^2 \mu_0 \varepsilon_0 \mu_{ru} \varepsilon_{ru} \tag{4}$$

where each layer (u) is denoted by its proper thickness (t_u) and by the EM intrinsic wave impedance (η_{iu}) of the constituent material at the ith value of frequency (f). The intrinsic wave impedance and the wave number (k_{iu}) are linked to the relative electric permittivity and magnetic permeability of the material constituting the uth layer (ε_{ru} and μ_{ru}, respectively) and to the vacuum electric permittivity and magnetic permeability (ε_0 and μ_0, respectively).

The adopted particle swarm optimization (PSO) algorithm evolves iteratively by minimizing the following objective function (OF):

$$OF = \sum_{frequency} \frac{|S_{RC} - T_{RC}|^2}{|S_{RC} + T_{RC}|^2} \tag{5}$$

where T_{RC} is the target RC, and S_{RC} is the simulated RC of a given multilayer—that is, a "particle" filling the space of solutions—obtained by combining Eqs. (1) and (2). Input of the design/optimization code is the target RC profile to fit and the electric permittivity of the materials in the database (the magnetic permeability is here neglected since the measurements of the dielectric parameters have shown that μ_r is unitary for all the materials under test). The output is the composition of the multilayered structure, that is, number of layers, layering sequence, and material/thickness of each layer. The PSO ranges within the space of solutions in order to minimize the OF defined in Eq.(4): the constraints are mainly imposed by the freedom degrees in exploring the database of available materials, in order to select the most appropriate for each layer, as well as the related thickness, limiting to a maximum of ten-layer structures. Further details on the EM wave/multilayer interaction modeling with a full description of the developed PSO algorithm can be found in some recent publications [21–23].

3. Results

The results of CEMA solutions for several RC targets are listed in **Tables 2–5**, where the target parameter, the materials of each layer coded with a number referred to **Table 1**, the thickness of each layer and the total thickness of the multilayer are reported. All the solutions are represented as ten-layer sequences: of course, a zero thick layer means that the PSO algorithm optimized the final layered material without the necessity of using that particular layer. In some cases, the indication of the PSO parameter α is also provided [21–23], whose meaning is here briefly recalled: by this parameter, the OF optimization is weighted with respect to the multilayer total thickness (e.g., $\alpha = 1$ means that only OF will be taken into account by PSO, without taking care of the thickness minimization). In the next **Figures 6–15**, the plot of the RC simulations is reported, comparing to the corresponding targets (red dashed lines). Almost all the simulations are based on RC target imitation by providing a PEC at the end of the layered material, except two. In particular, CEMA 02 has been designed providing free-space environment on both sides, that is, without the presence of PEC as the back-end of the structure. For CEMA 24, two different design scenarios have been taken into account: the first one represents the usual RC target imitation with a PEC-backed multilayer, while the second one explores the possibility to consider the TC as target parameter to be assigned to the PSO algorithm, thus requiring free space at both sides (here the other PSO parameter γ is introduced, to point out the weight in RC/TC addressing: $\gamma = 0$ indicates the RC optimization with no regard to TC, vice versa for $\gamma = 1$, intermediate values represent an optimization process that takes into account for TC and RC imitation simultaneously—cfr. CEMA02 plot). In this special case, the expected EM losses phenomena within the material increase with the frequency, thus progressively lowering down the TC values: such unwanted collateral effect is unavoidable since the higher the frequency the greater the skin-depth effect, which in turns increases the EM wave losses within thick materials.

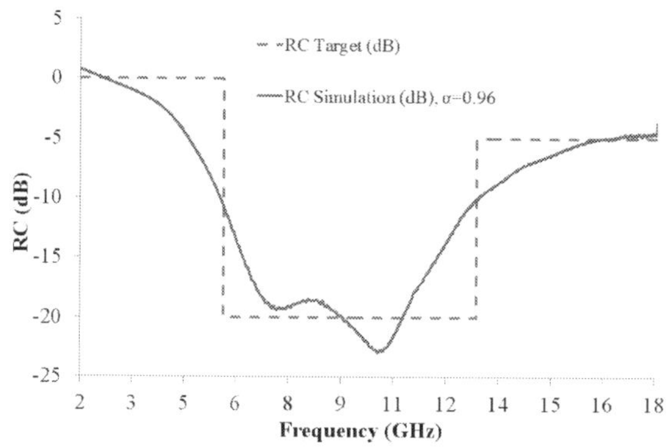

Figure 6. Plot of CEMA 00.

Target	CEMA	Materials	Layers thickness (mm)	Total thickness (mm)
T_{RC} 00	CEMA 00	18	4.5966	6.7069
		6	0	
		6	1.3962	
		2	0.4814	
		7	0.2162	
		10	0	
		18	0	
		13	0	
		5	0.0166	
		9	0	
		PEC	–	
T_{RC} 02	CEMA 02	17	0.6711	41.5812
		18	2.1669	
		12	3.4105	
		10	0	
		8	9.0000	
		19	1.1553	
		10	9.0000	
		13	0	
		7	9.0000	
		17	7.1774	
		Free space	–	
T_{RC} 13	CEMA 13 ($\alpha = 0.96$)	15	2.6898	44.1832
		11	1.2257	
		9	0.0798	
		11	9.0000	
		8	3.9273	
		3	8.2297	
		10	8.1644	
		12	9.0000	
		10	1.4261	

Target	CEMA	Materials	Layers thickness (mm)	Total thickness (mm)
		3	0.4403	
		PEC	–	
T_{RC} 13	CEMA 13 (α = 0.99)	11	0.0603	48.0260
		7	2.4181	
		12	9.0000	
		16	4.4037	
		7	0	
		11	0.6423	
		12	9.0000	
		15	9.0000	
		12	9.0000	
		12	4.5016	
		PEC	–	

Table 2. CEMA solutions by PSO for different target: layers sequence (materials coded as in **Table 1**), single layers, and total multilayer thickness.

Target	CEMA	Materials	Layers thickness (mm)	Total thickness (mm)
T_{RC} 15	CEMA 15 (α = 0.99)	1	8.5696	44.8402
		11	0	
		9	9.0000	
		18	4.9133	
		10	8.7613	
		12	0.6217	
		11	0.5130	
		6	9.0000	
		12	3.4488	
		17	0.0125	
		PEC	–	
T_{RC} 15	CEMA 15 (α = 0.96)	21	0	30.0380
		1	8.6290	
		12	0	

Target	CEMA	Materials	Layers thickness (mm)	Total thickness (mm)
		11	0	
		8	0.0046	
		8	0.4763	
		10	7.8559	
		10	0	
		16	4.7368	
		2	8.3354	
		PEC	–	
T_{RC} 16	CEMA 16 (α = 0.96)	19	9.0000	16.7440
		12	0	
		9	0	
		15	2.9726	
		5	2.0314	
		13	0	
		10	2.7400	
		13	0	
		16	0	
		14	0	
		PEC	–	
T_{RC} 16	CEMA 16 (α = 0.99)	15	9.0000	46.9286
		6	0	
		22	3.2445	
		12	1.3837	
		15	8.8354	
		12	9.0000	
		10	9.0000	
		12	3.3121	
		15	0	
		17	3.1529	
		PEC	–	

Table 3. CEMA solutions by PSO for different target: layers sequence (materials coded as in **Table 1**), single layers, and total multilayer thickness.

Target	CEMA	Materials	Layers thickness (mm)	Total thickness (mm)
T_{RC} 17	CEMA 17 (α = 0.99)	18	6.7023	52.3048
		8	4.4908	
		11	0	
		12	9.0000	
		10	0.5107	
		10	9.0000	
		12	1.1707	
		11	3.4329	
		6	9.0000	
		14	8.9974	
		PEC	–	
T_{RC} 17	CEMA 17 (α = 0.96)	18	5.7320	37.2708
		5	0	
		7	1.5600	
		5	0	
		10	2.9788	
		4	9.0000	
		10	0	
		7	0	
		14	9.0000	
		14	9.0000	
		PEC	–	
T_{RC} 20	CEMA 20 (α = 0.96)	18	6.8693	33.4005
		2	0	
		10	4.4797	
		3	6.7434	
		14	3.1642	
		14	9.0000	

Target	CEMA	Materials	Layers thickness (mm)	Total thickness (mm)
		6	1.3522	
		11	0	
		13	0	
		21	1.7917	
		PEC	–	
T_{RC}20	CEMA 20 (α = 0.99)	18	2.5959	44.5547
		15	4.1778	
		8	0.0476	
		8	0	
		12	0.3846	
		10	3.8949	
		11	9.0000	
		11	9.0000	
		12	8.9970	
		4	6.4568	
		PEC	–	

Table 4. CEMA solutions by PSO for different target: layers sequence (materials coded as in **Table 1**), single layers, and total multilayer thickness.

Target	CEMA	Materials	Layers thickness (mm)	Total thickness (mm)
T_{RC}23	CEMA 23 (α = 0.99)	13	9.0000	82.1359
		18	9.0000	
		21	7.1326	
		18	6.4060	
		13	9.0000	
		17	7.8792	
		18	7.8883	
		13	8.8366	
		13	9.0000	

Target	CEMA	Materials	Layers thickness (mm)	Total thickness (mm)
		13	7.9931	
		PEC	–	
T_{RC} 24	CEMA 24 only TC (α = 0.99, γ = 0)	20	8.5429	25.5742
		13	0	
		8	4.9744	
		14	9.0000	
		6	2.0820	
		14	0	
		9	0	
		19	0.9749	
		13	0	
		18	0	
		Free-space	–	
T_{TC} 24	CEMA 24 only RC (α = 0.99, γ = 1)	20	9.0000	48.9902
		22	9.0000	
		8	9.0000	
		9	0.1918	
		4	1.8001	
		12	9.0000	
		13	0	
		13	9.0000	
		13	1.9983	
		12	0	
		PEC	–	

Table 5. CEMA solutions by PSO for different target: layers sequence (materials coded as in **Table 1**), single layers, and total multilayer thickness.

The PSO algorithm shows an intriguingly effective mimic capability in almost all the cases, even for the most demanding (actually "not physical") targets conceived. Of course, the reliability degree is strictly linked to the target shape, that is, to the peaks number and depth as well as to the sharpness of the RC variations to be hunted. T_{RC} 00 and T_{RC} 02 present a simple filter-shaped profile: the corresponding solutions are able to broadly follow the target trend, even if the sharp (ideal) square-like behavior, with deep peaks up to 20 dB, does not allow a satisfactory imitation around the edges. In particular, CEMA 02 almost loses track of the target at the frequency range boundaries, probably due to the TC weighting in this free-space backed situation. Other regular square-like target profiles (multi-filter behavior) with RC oscilla-

http://dx.doi.org/10.5772/64213

tions of about 15 dB confirm both potentialities and issues of the method. In particular, the very difficult task of T_{RC} 23 imitation is actually impossible to tackle, even if the quasi-total reproduction of all the oscillating peaks by CEMA 23 represents a surprising result in this case. T_{RC} 16 is much easier to be imitated due to the greater peaks width, even if some inaccuracies are discovered for both the CEMA 16 solutions at lowest frequencies. The results obtained by CEMA 24 can be seen as intermediate case of the latter two. An excellent mimic effectiveness is provided by CEMAs 15, 17, and 20, which are able to closely follow the corresponding targets: in such cases, the algorithm is "aided" by the lower peaks depth and mainly by smoother RC variations in frequency. The difference in pursuing triangle-shaped *versus* square-shaped RC oscillations is well established by the two CEMA 13 solutions, which alternatively match and miss the exotic "stair-profile" of T_{RC} 13. For the T_{RC} 13, 15, 16, 17, and 20, two CEMA solutions have been proposed by slight variation of α. It can be noticed that at lower α (0.96) the algorithm finds solutions with thinner total thickness: that is reflected by the simulations, which shows a little bit more inaccuracy (black dashed lines) around the sharpest RC variations. Concerning the transmission targets ($\gamma \neq 0$), difficulty to conceive appreciable solutions in such framework due to the abovementioned intrinsic skin-depth effects, evident in both the situations investigated, needs to be pointed out. In T_{RC} 02, the parameter γ is set to 0.5 (i.e., the optimization of RC and TC is equally weighted) and a constant low transmission rate is defined: besides the loss effects, the simultaneous research of a layering to follow the RC target leads to an intermediate solution, not suitable for the required transmission (otherwise a quasi-full absorbing structure would have been identified). As for the T_{TC} 24 ($\gamma = 1$), a challenging square-shaped transmission profile is defined (by analogy with T_{RC} 24, $\gamma = 0$). The quite fast target oscillations are only remotely hinted by the CEMA 24 solution, whose average TC substantially decreases at higher frequency; nevertheless, it is worth noticing that the corresponding reflection presents a peaks series roughly reversed, in respect to the transmission target.

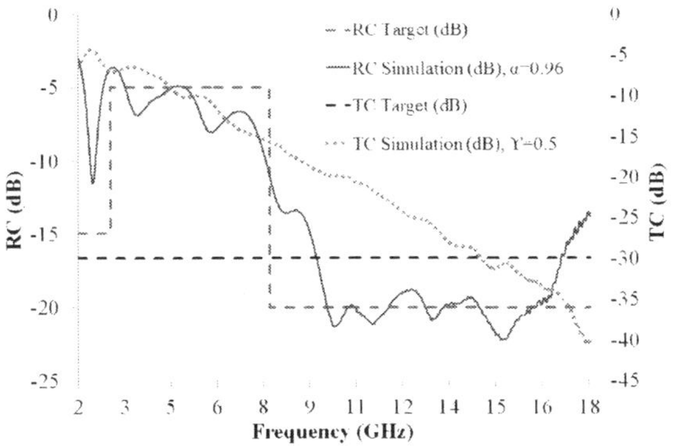

Figure 7. Plot of CEMA 02.

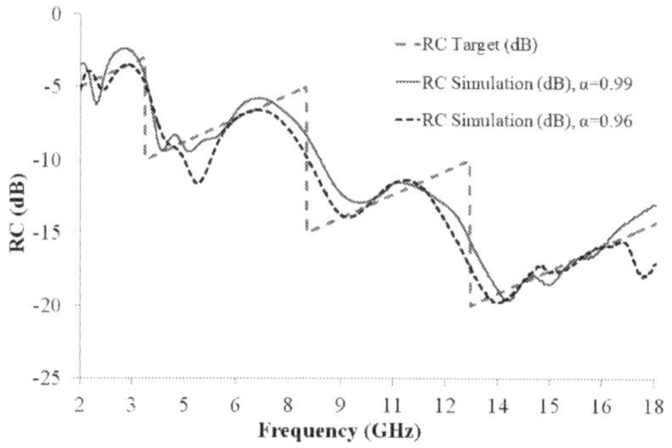

Figure 8. Plot of CEMA 13.

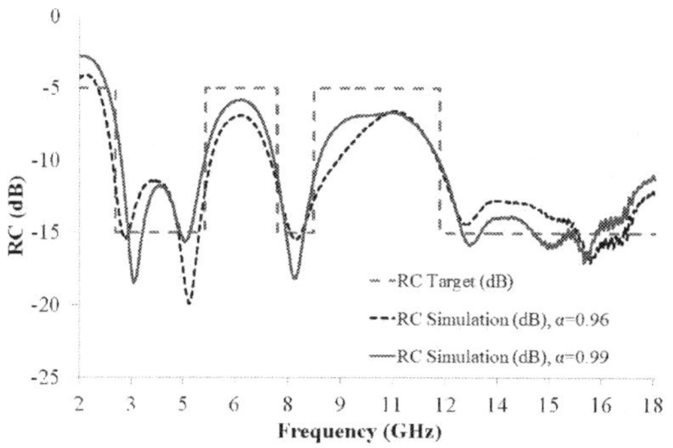

Figure 9. Plot of CEMA 15.

A promising level of confidence is suggested from the reported results in each phase of the work, from the database materials preliminary characterization to the mathematical modeling of the EM field/matter interaction, up to the PSO algorithm design optimization. The capability of the layering approach lies in the frequency selective interaction between the different materials and the EM field. Such concept is well explained by a finite element method (FEM) simulation where CEMA13 (α = 0.96) has been deeply analyzed by studying the multilayer structure inserted in a 50 Ω coaxial airline of the same type used to measure the dielectric properties of the database materials. The FEM analysis shows how the EM field propagates through the several layers at two different frequencies, 2.5 and 14 GHz, where the reflection coefficient, respectively, results maximum and minimum. Commercial code COMSOL Multiphysics has been used for the FEM analysis [20, 24]. In **Figure 16**, the mesh of the multilayer within the coaxial waveguide is depicted: the blue part on the left is the empty section of the airline, while the other is the part filled with ten layers of nanocomposite materials. The airline is simulated using a perfect electric conductor for inner, outer, and back

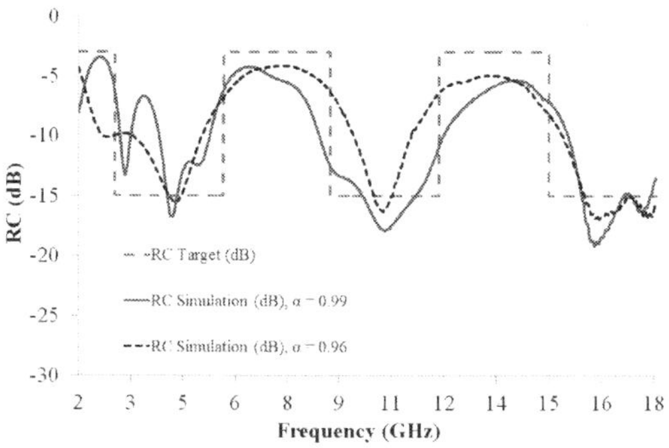

Figure 10. Plot of CEMA 16.

surfaces. The Port1 is placed at the beginning of the coaxial airline on the left side of the figure. The maximum element size of the mesh is 1/10 of the minimum EM wavelength ($\lambda = 3 \times 10^8/f_{max} \sim 1.6$ cm), while the minimum element size of the mesh is in the order of 10^{-7} m: these dimensions enable an optimal growth of the mesh and guarantee the needed accuracy in the EM analysis. In the coaxial airline, Port 1 is used to compute the scattering parameter S_{11}, related to the reflection coefficient. In the numerical simulation, the power set at the port is 3 dBm (2 mW) as in the real measurement. All the parameters adopted to run the FEM simulation are chosen in such a way to reproduce the experimental conditions of the microwave characterization. In particular, the real and imaginary part of the complex electrical permittivity measured over the 2–18 GHz frequency band to create the database of nanocomposite materials have been adopted: such values are interpolated by COMSOL code in order to get out a functional relationship between dielectric properties and frequency. The EM interfaces of the code cover the modeling of EM fields and waves in the frequency domain, formulating and solving the differential form of Maxwell's equations together with the initial and boundary conditions.

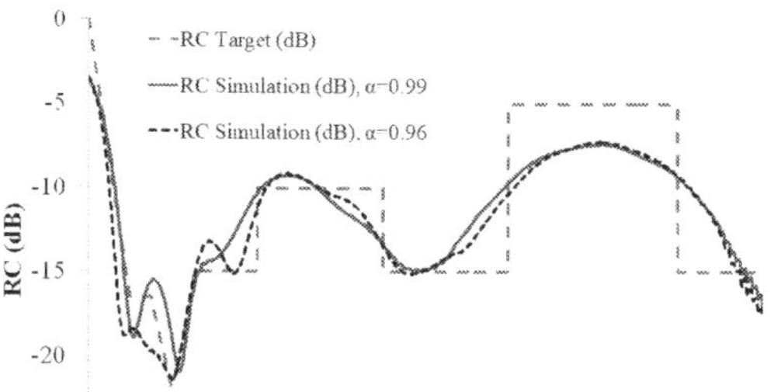

Figure 11. Plot of CEMA 17.

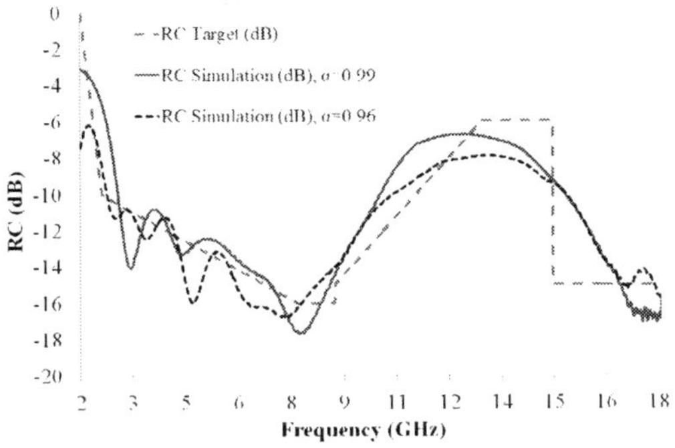

Figure 12. Plot of CEMA 20.

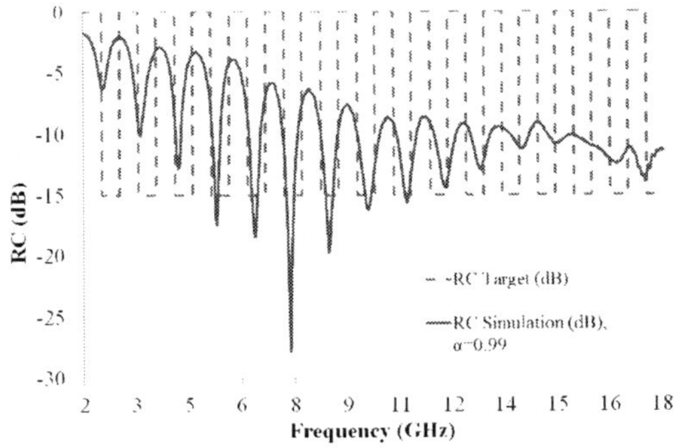

Figure 13. Plot of CEMA 23.

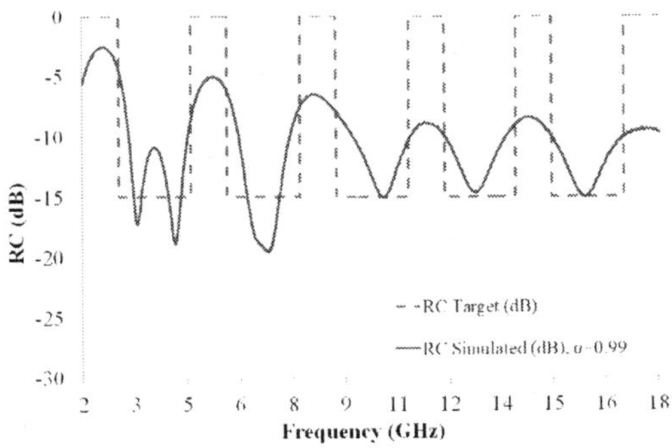

Figure 14. Plot of CEMA 24, optimization of RC.

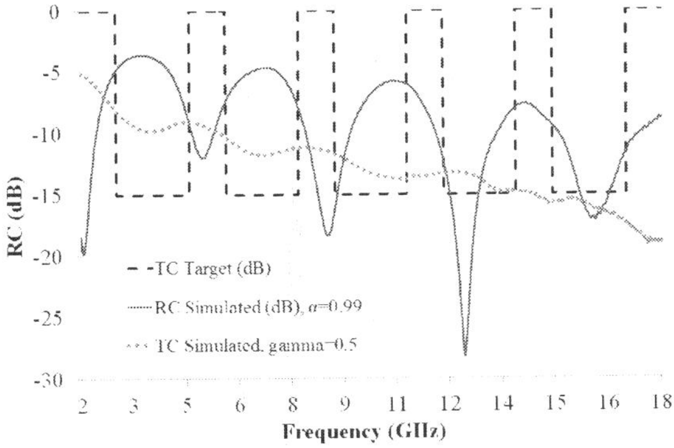

Figure 15. Plot of CEMA 24, optimization of TC.

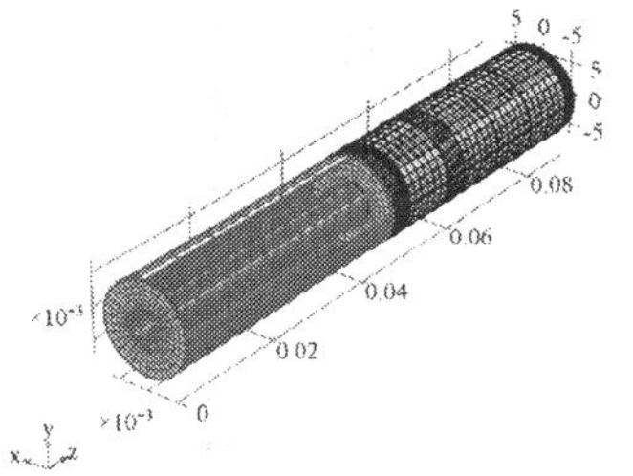

Figure 16. Mesh of the CEMA 13 (α = 0.96), ten-layered structure in the coaxial airline setup.

In **Figures 17** and **18**, the electric field (V/m) is reported at 2.5 and 14 GHz respectively. In **Figures 19** and **20**, the current density (A/m^2) arrow lines and magnetic field (A/m) circular lines for the ten layers of CEMA13 are shown (arrow lines are used to show the region where the highest EM losses occur along the structure). In **Figures 21** and **22**, the power loss density (W/m^3) at 2.5 and 14 GHz, respectively, is shown. At 14 GHz, the multilayer design provides the impedance matching with free space, thus resulting in a higher level of EM absorption, as compared to the values of the electric field at 2.5 GHz. Alongside, the significant current density at 14 GHz is well visible in the first and second layers, whereas at 2.5 GHz the impedance mismatch causes a great reflection of the EM field at the first layer, thus determining a lower power loss within the deep bulk. In fact, the power loss density at 2.5 GHz is quite lower compared to the value obtained at 14 GHz, where geometrically power losses are mainly confined within the 5 first layers. In other words, at 2.5 GHz, the EM field propagates through all the layers due to the impedance mismatch end. Due to the higher value of the reflection coefficient, the majority of the power is reflected back at the first air-multilayer interface. At 14 GHz, the multilayer is impedance matched, the reflection coefficient is quite lower and the absorbed power dramatically increases. In **Figure 23**, a comparison between the RC Target, RC simulated with mathematical model and RC simulated by FEM analysis is shown. The discrepancies with FEM simulation are due to the mesh and the approximation of the EM field solution, which is more difficult at high frequencies.

Figure 17. Electric field (V/m) at 2.5 GHz.

Figure 18. Electric field (V/m) at 14 GHz.

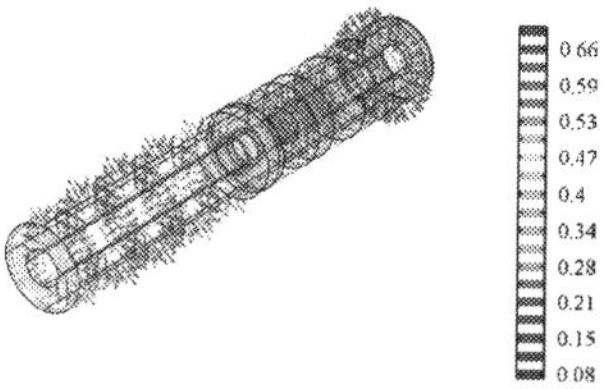

Figure 19. Magnetic field (A/m) circular line and current density (A/m²); red arrow lines at 2.5 GHz.

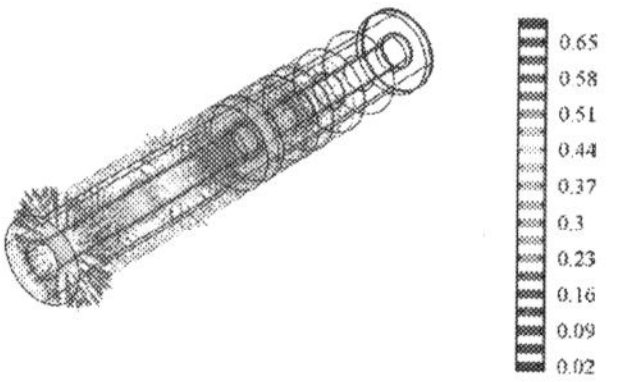

Figure 20. Magnetic field (A/m) circular line and current density (A/m²); red arrow lines at 14 GHz.

Figure 21. Power loss density (W/m³) at 2.5 GHz.

Figure 22. Power loss density (W/m³) at 14 GHz.

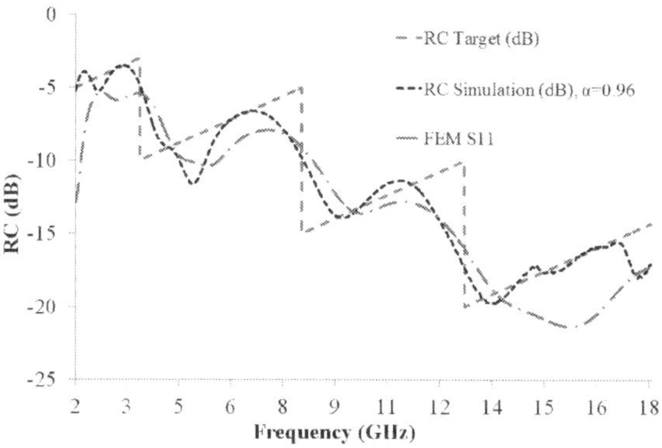

Figure 23. Comparison of RC Target, RC simulated, and RC simulated with FEM.

4. Conclusions

In this chapter, an attempt to identify a novel approach to design and optimize EM reference materials for metrological applications was introduced. The results highlight the intriguing potentialities of the proposed strategy, since even hard microwave reflection target profiles are successfully addressed by the numerical design optimization technique developed. It is clear that several upgrades can be introduced in order to better refine the process. The above mentioned examples highlight some root of inaccuracy, especially when quite sharp edges occur along the target profile. Of course, the harder it is to follow a given profile, the more complex the optimized structure will become out in terms of layers number and thinness. Although the intriguing capability of the PSO algorithm to find solutions fitting such challenging targets, it has to be outlined that the corresponding CEMA solutions are composed by at least four layers (almost always six or more), with total thickness over 4 cm, in the most of the situations. Furthermore, quite often the layers that fulfill the crucial EM absorbing role have thickness well below 1mm, thus pointing out that an eventual production process could represent a big task in this framework. Beyond possible PSO algorithm enhancements (convergence parameters, particles population, iterations), the effectiveness of the proposed strategy in terms of mimic capability would be significantly improved by enriching the materials database. The possibility to draw on a "quasi-continuous" spectrum of dielectric properties should allow the PSO algorithm to find solutions for reproducing any RC profile even more faithfully. On the other hand, by proper constraints in terms of layers number and thickness, the increased availability of materials should make the algorithm able to identify multilayered combinations easy to be practically achieved. In this respect, the employment of composites reinforced with nanoparticles sounds as the right way forward. With respect to the conventional materials, in fact, the family of nano-filled composites allows the unique opportunity to finely adjust the EM absorption property just by tuning the weight percentage of the nanofiller within the matrix. At this purpose, the lower the incremental step in weight percentage, the greater the possibility of fine tuning: thus, the greater the database population, the better the possibility to select appropriate materials for mimic purposes. The research development will thus be addressed to enrich the database population, in order to obtain even more precise and feasible solutions for the widest variety of electromagnetic requirements.

Author details

Davide Micheli*, Roberto Pastore, Antonio Vricella and Mario Marchetti

*Address all correspondence to: davide.micheli@uniroma1.it

Astronautic Electric and Energy Engineering Department, Sapienza University of Rome, Rome, Italy

References

[1] Sheffield RG, The official F-19 stealth fighter handbook, Radnor, Pennsylvania: Pam Williams ed, 1989. ISBN 10: 0874552176: 1–184.

[2] Knott EF, Shaeffer J, Tuley M, Radar cross section, 2nd edn, SciTech Publishing Inc., 2004: 112–115.

[3] Mosallaei H, Rahmat-Samii Y. RCS reduction of canonical targets using genetic algorithm synthesized RAM. IEEE Trans. Antennas Propag. 2000; 48(10): 1594–1606. doi:10.1109/8.899676

[4] Vinoy KJ, Jha RM, Radar absorbing materials—from theory to design and characterization. Boston: Kluwer, 1996. ISBN 978-1-4613-8065-8: 1–173.

[5] Sagalianov IY, Vovchenko LL, Matzui LY, Lazarenko AA, Oliynyk VV, Lozitsky OV, Ritter U. Optimization of multilayer electromagnetic shields: A genetic algorithm approach. Mat.-wiss. u. Werkstofftech. 2016. doi:10.1002/mawe.201600483.

[6] Micheli D, Apollo C, Pastore R, Marchetti M. X-band microwave characterization of carbon-based nanocomposite material, absorption capability comparison and RAS design simulation. Composites Sci. Technol. 2010; 70(2): 400–409. doi:10.1016/j.compscitech.2009.11.015

[7] Li N, Huang Y, Du F, He X, Lin X, Gao H, Ma Y, Li F, Chen Y, Eklund PC. Electromagnetic interference (EMI) shielding of single-walled carbon nanotube epoxy composites. Nano Lett. 2006; 6(6): 1141–1145. doi:10.1021/nl0602589

[8] Chang CM, Chiu JC, Jou WS, Wu TL, Cheng WH. New package scheme of a 2.5-Gb/s plastic transceiver module employing multiwall nanotubes for low electromagnetic interference. IEEE J. Select. Top. Quantum Electron. 1989; 25(5): 1025–1031. doi:10.1109/JSTQE.2006.879534

[9] Bogush V, Borbotko T, Kolbun N, Lynkov L. Novel composite shielding materials for suppression of microwave radiation. In: International conference on microwaves radar wireless communications, 2006. MIKON 2006. 645–647. doi:10.1109/MIKON.2006.4345262

[10] Emerson W. Electromagnetic wave absorbers and anechoic chambers through the years. IEEE Trans. Antennas Propag. 1973; AP-21(4): 484–490. doi:10.1109/TAP.1973.1140517

[11] Holloway CL, DeLyser RR, German RF, McKenna P, Kanda M. Comparison of electromagnetic absorber used in anechoic and semi-anechoic chambers for emissions and immunity testing of digital devices. IEEE Trans. Electromagn. Compat. 1997; 39(1): 33–47. doi:10.1109/15.554693

[12] Davide Micheli, Roberto Pastore, Gabriele Gradoni and Mario Marchetti. Tunable nanostructured composite with built-in metallic wire-grid electrode. AIP Adv. 2013; 2158–3226/3(11): 112132–112137. doi:10.1063/1.4837916

[13] Meissner T, Wentz FJ. The complex dielectric constant of pure and sea water from microwave satellite observations. IEEE T. Geosci. Remote 2004. 42, 1836–1849. doi:10.1109/TGRS.2004.831888

[14] Hoeben R, Troch PA. Assimilation of active microwave observation data for soil moisture profile estimation. Water Resour. Res. 2000; 36, 2805–2819. doi: 10.1029/2000WR900100

[15] Martinez-Lorenzo JA, Rappaport C, Sullivan R, Angell A. Standoff concealed explosives detection using millimeter-wave radar to sense surface shape anomalies. Antennas and Propagation Society International Symposium, 2008. AP-S 2008. IEEE. pp 1–4. doi:10.1109/APS.2008.4619051

[16] Robert J, Douglass J, Gorman D, Burns TJ. System and method for standoff detection of human carried explosives. Patent US 7800527 B2, PCT/US2005/036593, 2010.

[17] Micheli D, Apollo C, Pastore R, Marchetti M. Modeling of microwave absorbing structure using winning particle optimization applied on electrically conductive nanostructured composite material. 19th ICEM 2010, Rome (IT) 2010; ISBN 978-1-4244-4174-7: 1–10. doi:10.1109/ICELMACH.2010.5607881

[18] Micheli D, Pastore R, Marchetti M, Gradoni G, Moglie F, Mariani Primiani V. Modeling and measuring of microwave absorbing and shielding nanostructured materials. IEEE EMC Europe 2012, Rome (IT) 2012; ISBN 978-1-4673-0718-5: 1–5. doi:10.1109/EMCEurope.2012.6396810

[19] Micheli D, Radar absorbing materials and microwave shielding structure design, LAP Lambert Academic Publishing 2012; ISBN 978-3-8465-5939-0: 334–446.

[20] Micheli D, Vricella A, Pastore R, Marchetti M. Synthesis and electromagnetic characterization of frequency selective radar absorbing materials using carbon nanopowders. Carbon 2014; 77: 756–774. doi:10.1016/j.carbon.2014.05.080

[21] Micheli D, Pastore R, Gradoni G, Mariani Primiani V, Moglie F, Marchetti M. Reduction of satellite electromagnetic scattering by carbon nanostructured multilayers. Acta Astron 2013; 88: 61–73. doi:10.1016/j.actaastro.2013.03.003

[22] Micheli D, Pastore R, Apollo C, Marchetti M, Gradoni G, Mariani Primiani V, Moglie F. Broadband electromagnetic absorbers using carbon nanostructure-based composites. IEEE Trans. Microwave Theory Tech. 2011; 59(10): 2633–2646. doi:10.1109/TMTT.2011.2160198

[23] Micheli D, Apollo C, Pastore R, Barbera D, Bueno Morles R, Marchetti M, Gradoni G, Mariani Primiani V, Moglie F. Optimization of multilayer shields made of composite nanostructured materials. IEEE Trans. Electromagn. C 2012; 54(1): 60–69. doi:10.1109/TEMC.2011.2171688

[24] Pryor RW. Multiphysics modeling using COMSOL: a first principles approach. Sudbury: Jones & Bartlett Learning; 2009.